Modeling and Simulation of Fluidized Bed Reactors for Chemical Looping Combustion

Ramesh K. Agarwal • Yali Shao

Modeling and Simulation of Fluidized Bed Reactors for Chemical Looping Combustion

 Springer

Ramesh K. Agarwal
Mechanical Engineering
and Materials Science
Washington University in St. Louis
Saint Louis, MO, USA

Yali Shao
Energy and Mechanical Engineering
Nanjing Normal University
Nanjing, China

ISBN 978-3-031-11337-6   ISBN 978-3-031-11335-2   (eBook)
https://doi.org/10.1007/978-3-031-11335-2

This Springer imprint is published by the registered company Springer Nature Switzerland AG
The registered company address is: Gewerbestrasse 11, 6330 Cham, Switzerland

Paper in this product is recyclable.

Preface

Carbon dioxide released from the burning of fossil fuels is largely responsible for the global warming and climate change. The carbon capture, utilization, and storage (CCUS) technology is being developed as an effective method to mitigate CO_2 emissions. In the last two decades, chemical looping combustion (CLC) has emerged as one of the most effective carbon capture utilization and sequestration (CCUS) approaches, which has gained sustained interest by the energy generation industry based on fossil fuels due to the inherent capability of CLC in almost pure CO_2 capture. While the CLC technology is still under development and awaits industrial-scale demonstration, hundreds of research centers and universities worldwide have devoted a great deal of effort to the exploration of this technology. With the availability of high-speed computers and commercial software, increasing attention has been paid to the analysis and computation of the flow field and reaction characteristics of the complex multi-phase flow in CLC. The application of a numerical approach to the CLC process started in 2008 with the first investigation of flow dynamics in a bubbling fluidized-bed fuel reactor. Since then, an enormous amount of research has been conducted using different types of fuels, oxygen carriers and reactors.

Even though many papers on CLC simulations have been published, the systematic introduction and summary of numerical modeling and simulation of the CLC process are still lacking, which increases the difficulty for the comprehensive understanding and quick grasp of numerical simulation methods and concepts. With the authors' rich experience in the fields of computational fluid dynamics (CFD), CCUS, and combustion, they beautifully summarize the state of the art in numerical simulation of the CLC process, synthesizing and summarizing large amount of previous work in the literature.

The core of the book is aimed at addressing the need of scientific/engineering community for summary of the latest research and state of the art in the field of modeling and simulations of chemical looping combustion processes; it introduces fundamental concepts and methodologies followed by specific examples. This book is first of its kind in the literature. The unique feature of the book is the emphasis on comprehensive review of almost all the typical physical and simulation processes, including CFD simulations of CLC. Overall, this book:

- Enables the readers to quickly and easily understand the principle of chemical looping and its applications by following the well-organized content.
- Offers detailed and easy-to-follow guidelines to conduct the process simulation and CFD simulation of various chemical looping processes in different reactor configurations.
- Serves a wide range of readers including undergraduate/postgraduate students with educational background in chemical, mechanical, or environmental engineering, as well as researchers and engineers who are trained in similar background and are active in research in CLC, CCUS, and related technologies.

The following is a summary of the book contents: Chaps. 1 and 2 of the book include the basic concepts of chemical looping combustion, including reverse CLC, description of various types of fluidized-bed reactors as well as the oxygen carriers used in CLC reactors. Chapter 3 demonstrates the application of process simulation software Aspen Plus to in-situ gasification chemical looping combustion (iG-CLC) and chemical looping with oxygen uncoupling (CLOU) processes as well as single and multistage reactor design. Chapter 4 provides the description of governing equations in Eulerian and Lagrangian framework for gas-solid multi-phase flows. Chapters 5–11 discuss how different numerical approaches are used for simulations of various fluidized-bed fuel reactors, single-loop/dual-loop systems, carbon strippers, gas/solid fuels, etc. Chapter 12 presents the application of CFD simulation for scale-up of the CLC reactor and assesses different scaling methodologies. Chapter 13 introduces the frequently used machine learning methods and their application in CLC systems. Chapter 14 describes some concepts and applications of chemical looping beyond combustion. Chapter 15 describes how carbon capture and sequestration can be combined with CLC and the application of CFD technology for the simulation of CO_2 sequestration in two large-scale, identified deep saline aquifers. In summary, the book provides information on the implementation of state-of-the-art modeling, numerical simulation approaches, and methodology for chemical looping. The authors hope that the book will be beneficial and become a go-to source to students, academic researchers, and industry practitioners interested in chemical looping technology and will also help them in making contributions toward the further development of CLC technology.

Many people have contributed toward the material presented in this book. The authors wish to express their sincere appreciation to the students of Prof. Agarwal: Subhodeep Banerjee, Chris Chivetta, Spencer Chen, Wei Dai, Kartik Deshpande, Justin Lam, Guanglei Ma, William Meng, Suraj Puuvada, Pravin Sivabalan, Hongming Sun, Mengqiao Yang, Xiao Zhang, and Zheming Zhang who worked with him at Washington university in Saint Louis for their valuable contributions to perform some of the process simulations using ASPEN Plus and CFD simulations presented in this book. The authors would like to acknowledge the contributions of

Prof. Ling Zhou, Prof. Ling Bai, and Prof. Weidong Shi of Jiangsu University in China who have collaborated with Prof. Agarwal for many years in modeling and simulation of CLC. The contributions and collaboration of Prof. Xudong Wang of Nanjing Technical University and Prof. Baosheng Jin of Southeast University are also gratefully acknowledged.

The authors are thankful for some initial financial support provided by the Consortium for Clean Coal Utilization (CCU) at Washington University in St. Louis for conducting some of the research reported in this book. Lastly but not the least, it has also been a great pleasure working with the team at Springer in publishing this book.

Finally, the authors want to express their gratitude to their respective family members for their encouragement and unwavering support during the preparation of the book.

Saint Louis, MO, USA   Ramesh K. Agarwal
Nanjing, China   Yali Shao

Contents

Chapter 1
Introduction

In recent years, several technologies have been demonstrated to capture CO_2 emissions from fossil-fueled power plants and greatly reduce emissions into the atmosphere. These technologies can be broadly categorized as pre-combustion capture such as the integrated gas combined cycle (IGCC), post-combustion capture such as sorbent-based absorption, and oxy-fuel combustion. However, each of these technologies require a separate process to isolate CO_2 from the other gases, which consumes much of the total energy produced by the plant and can lead to a significant increase in the cost of electricity. One technology that has shown great promise for high-efficiency low-cost carbon capture is chemical looping combustion (CLC). The CLC process typically utilizes two reactors—an air reactor and a fuel reactor—and a metal oxide oxygen carrier that circulates between the two reactors, as illustrated in Fig. 1.1a. Another setup for CLC that has been documented in the literature employs a single vessel with a packed bed of oxygen carrier that is alternatingly used as an air and fuel reactor via a high-temperature gas switching system, shown in Fig. 1.1b. The primary advantage of CLC is that the combustion of fuel in the fuel reactor takes place in the absence of air using oxygen provided by the oxygen carrier; the flue stream from the fuel reactor is not contaminated or diluted by other gases such as nitrogen. This provides a high-purity carbon dioxide stream available for capture at the fuel reactor outlet without the need for an energy-expensive gas separation process. The reduced oxygen carrier from the fuel reactor is pneumatically transported to the air reactor where it is re-oxidized by oxygen from air and circulated back to the fuel reactor to complete the loop.

The only energy cost of separation associated with CLC is the cost of solid recirculation. This is considerably lower than the benchmark for pre-combustion technologies for carbon capture such as oxy-fuel combustion where the oxygen separation process can consume about 15% of the total energy. Therefore, CLC holds the answer as the next-generation combustion technology due to its potential to allow CO_2

R. K. Agarwal, Y. Shao, *Modeling and Simulation of Fluidized Bed Reactors for Chemical Looping Combustion*, https://doi.org/10.1007/978-3-031-11335-2_1

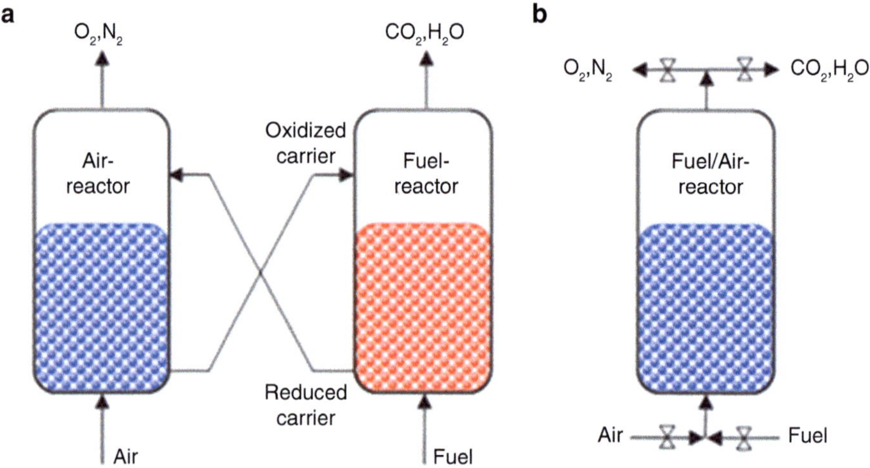

Fig. 1.1 Schematic representation of a chemical looping combustion system with (**a**) interconnected two reactors and (**b**) packed bed with alternating flow [1]

capture with little to no effect on the efficiency of the power plant. Several studies on the energy and exergy of CLC systems in the literature suggest that power efficiencies greater than 50% can be achieved along with nearly complete CO_2 capture [2–4].

References

1. H. Kruggel-Emden, F. Stepanek, A. Munjiza, A study on the role of reaction modeling in multi-phase CFD-based simulations of chemical looping combustion. Oil Gas Sci. Technol. d'IFP Energies Nouv. **66**(2), 313–331 (2011)
2. J. Wolf, M. Anheden, J. Yan, Performance analysis of combined cycles with chemical looping combustion for CO2 capture. In *Proceedings of 18th Pittsburg Coal Conference, December*; 2001; pp 3–7
3. J.L. Marion, N. Mohn, G.N. Liljedahl, N. Nsakala, J.X. Morin, P.P. Henriksen, Technology options for controlling CO2 emissions from fossil fueled power plants. In *Proceedings of 5th Annual Conference on Carbon Capture and Sequestration*, 2006
4. H.E. Andrus, G. Burns, J.H. Chiu, G.N. Liljedahl, P.T. Stromberg, P.R. Thibeault, Hybrid combustion-gasification chemical looping coal power technology development, Phase III–final report. *National Energy Technology Laboratory Albany, OR, Report No. PPL-08-CT-25* 2008

Chapter 2
Fundamental Concepts

2.1 Principles of Chemical Looping Combustion

In conventional solid-fueled combustion process, the solid fuel is first pyrolyzed to release volatile components and char which subsequently react with oxygen from the air. In the CLC process, the oxygen carrier takes the place of oxygen. The oxygen carrier not only provides the oxygen needed for combustion but can also cast as the catalyst for tar cracking and gasification of char. In the chemical looping combustion process, the conventional one-step reaction between the fuel and the air is divided into two steps occurring at two separate reactors, namely, a fuel reactor and an air reactor, as illustrated in Fig. 2.1. In the fuel reactor, the fuel is introduced and reacts with the solid oxygen carrier to form CO_2 and H_2O. Without the existence of N_2, high-concentration CO_2 can be obtained via simple condensation without much energy penalty at the exit of fuel reactor. In the air reactor, the air is introduced to react with the reduced oxygen carrier from the fuel reactor, and the exit flue gas contains N_2 and unreacted O_2. Generally, reaction in the fuel reactor is endothermic, and reaction in the air reactor is exothermic, but the total amount of heat released in the CLC process is the same as that from the conventional combustion. The solid oxygen carrier continuously circulates between the two reactors to not only transport the oxygen but also the heat from the air reactor to fuel reactor.

The CLC technology is initially used for the gas fuels such as CH_4, H_2, syngas, etc. When the solid fuel is used, the reaction rate between solid fuel and solid oxygen carrier is rather slow, which is the key factor limiting the reaction process. In order to enhance combustion efficiency, various types of CLC technologies have been developed for solid fuels, mainly including in situ gasification chemical looping combustion (iG-CLC) and chemical looping with oxygen uncoupling (CLOU), as illustrated in Fig. 2.2.

R. K. Agarwal, Y. Shao, *Modeling and Simulation of Fluidized Bed Reactors for Chemical Looping Combustion*, https://doi.org/10.1007/978-3-031-11335-2_2

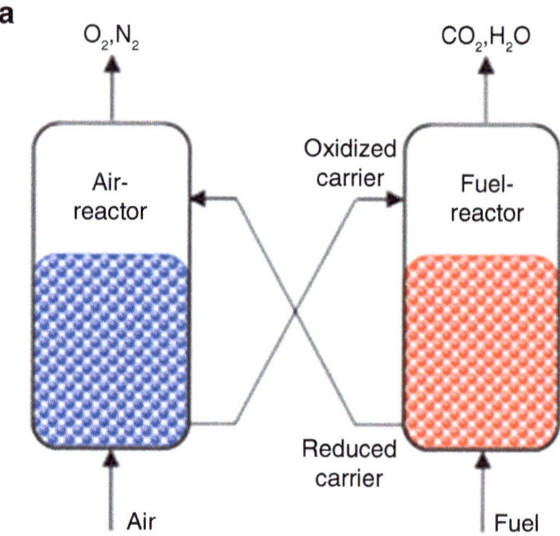

Fig. 2.1 Schematic diagram of chemical looping combustion technology [1]

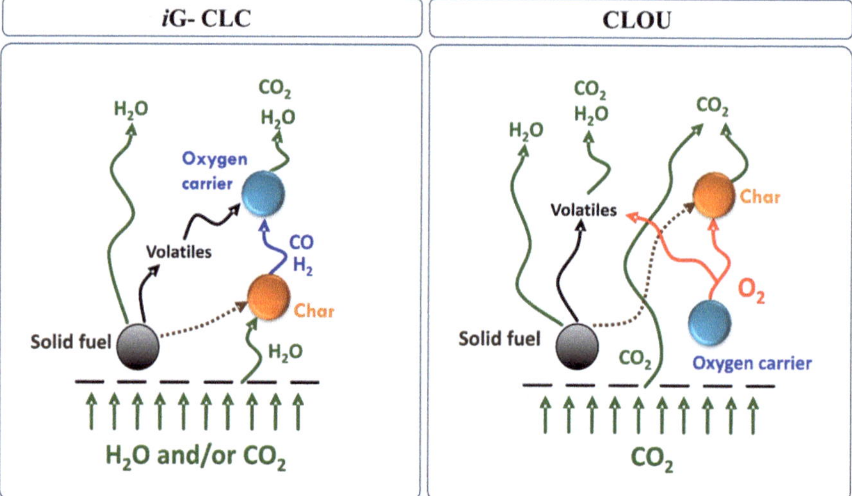

Fig. 2.2 Two types of CLC technologies used for solid fuel: (**a**) iG-CLC technology; (**b**) CLOU technology [2]

The iG-CLC technology integrates the gasification and combustion processes together. The overall reaction scheme is displayed in Eqs. 2.1–2.4, among which the gasification of char is the rate-controlling step. The gasification agent and oxygen carrier are introduced to the fuel reactor at the same time. The solid fuel such as coal is first pyrolyzed to char and gas volatiles including CO, H_2, CH_4, etc. at high temperature. The gas volatiles are oxidized by the oxygen carrier to CO_2 and H_2O, while the solid char is gasified by the gasification agent such as water vapor or CO_2 to form CO and H_2. In theory, the solid char can also react with oxygen carrier to release heat, but the reaction rate between solid particles is much slower than the gas-solid reaction. In spite of this, the gasification rate in the iG-CLC process is faster than the conventional gasification because of the high CO_2/H_2O concentration in the atmosphere. To realize full conversion of carbon in iG-CLC reactors, a carbon stripper located between the fuel reactor and air reactor is essential to separate the unreacted char particles from the oxygen carrier before the char particles reach the air reactor.

$$\text{Solid fuel} \rightarrow \text{Volatile matter} + \text{Char} \tag{2.1}$$

$$\text{Char (mainly carbon)} + H_2O \rightarrow CO + H_2 \tag{2.2}$$

$$\text{Char (mainly carbon)} + CO_2 \rightarrow CO \tag{2.3}$$

$$MeO_x + H_2, CO, \text{volatile matter} \rightarrow MeO_{x-1} + H_2O + CO_2 \tag{2.4}$$

To avoid the gasification step and intensify the reaction process, Mattison et al. [3] developed the CLOU technology by using a specific type of oxygen carrier such as CuO and Mn_2O_3. The lattice oxygen in the oxygen carrier can be released as molecular oxygen at high temperature, and hence the fuel does not have to react directly with the metal oxides. The CLOU technology also requires the oxygen carrier to have the ability to react with oxygen from air in the air reactor. The favorable reaction performance of CLOU process has been validated by much existing research [4–6]. In CLOU, the volatiles and char from coal directly mix and react with the molecular oxygen, as given by Eqs. 2.5–2.8. Without the gasification of coal, the gasification agent is not needed in the CLOU process. When comparing iG-CLC and CLOU, the reactor configurations are similar, while the main difference is the type of oxygen carrier used in the reactor:

$$\text{Solid fuel} \rightarrow \text{Volatile matter} + \text{Char} \tag{2.5}$$

$$MeO_x \rightarrow MeO_{x-2} + O_2 \tag{2.6}$$

$$\text{Char(mainly carbon)} + O_2 \rightarrow CO_2 \tag{2.7}$$

$$H_2, CO, \text{volatile matter} + O_2 \rightarrow CO_2 + H_2O \tag{2.8}$$

2.2   Fuel Reactor Designs

High performance in chemical looping process requires good contact between the oxygen carrier and the gas species as well as smooth solid circulation between two reactors. Given gas-solid contact and reaction efficiency is strongly related to reactor configuration, selection of reactor type is very critical [7]. Figure 2.3 shows the typical fluidized bed type for the fuel reactor, including bubbling fluidized bed (countercurrent gas-solid flow), spouted-fluidized bed (countercurrent gas-solid

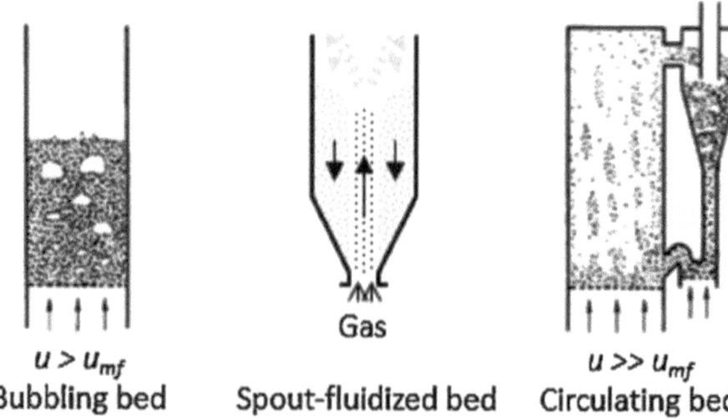

Fig. 2.3 Different types of fluidized bed fuel reactor for CLC [7]

flow), and circulating fluidized bed (cocurrent gas-solid flow) [8]. The bubbling fluidized bed is able to provide sufficient residence time for particles. However, the gas residence time is relatively short, and the existence of bubbles will provide shortcut for gases, which will adversely affect the gas-solid contact and reaction efficiency. Increasing the bed height to achieve longer gas residence time will inevitably cause larger pressure drop. The similar problem can be avoided in the circulating fluidized bed. The gas residence time can be prolonged via increasing the bed height without increasing the total pressure drop. In addition, no bubbles exist in the circulating fluidized bed riser, and the intense flow regime ensures sufficient gas-solid contact and reaction along the majority of the bed height. As to the spouted-fluidized bed, the particle spout forms, and particles fall back to the beds along the sides of the reactors, which greatly promotes the solid mixing in the reactor, thus enhancing the gas-solid reaction performance.

Besides, a moving bed was developed as the fuel reactor in Ohio State University, as shown in Fig. 2.4 [9]. In the moving bed, the oxygen carrier moves downward by gravity and the gas moves upward. The reactor is divided into three regions: (i) coal injection and devolatilization of coal, (ii) reduction of oxygen carrier by gaseous volatiles from coal in Sect. 2.1, and (iii) coal char gasification and the oxygen carrier reduction in Sect. 2.2. The coal is devolatilized very quickly once entering the high-temperature fuel reactor. The volatiles move upward and react with oxygen carrier in Sect. 2.1. The coal char moves downward to Sect. 2.2 and gasified by CO_2/H_2O, and then the gasification products react with the oxygen carrier.

To enhance the reaction efficiency, multistage fluidized bed fuel reactor is designed by connecting several independent fluidized beds via gas distributor, standpipe, multi-orifice plate, etc., which is able to effectively restrain the gas-solid back-mixing and prolong the gas-solid residence time. Thon et al. [10] designed a multistage fluidized bed fuel reactor based on two bubbling fluidized beds. The fresh oxygen carrier from the air reactor first moves to the top-bubbling fluidized bed and reacts with the combustible gas from the bottom bed. The partially reduced oxygen carriers further move to the bottom bubbling fluidized bed via a standpipe to react with the fuels. The intention of the two-stage fuel reactor design is to enhance the conversion of combustible gases released in the lower fuel reactor by contacting them with freshly regenerated oxygen carrier in the upper stage.

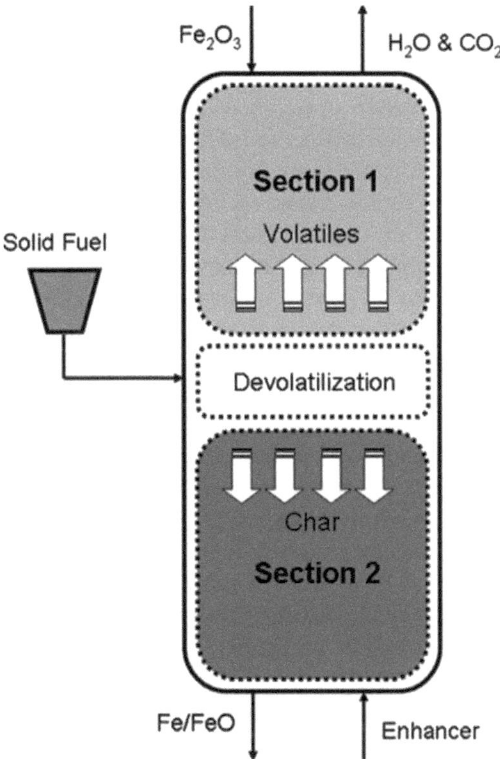

Fig. 2.4 Moving bed fuel reactor from Ohio State University [9]

Similarly, Shen et al. [11] also built up a multistage fuel reactor based on a bottom-spouted-fluidized bed and a top-bubbling fluidized bed. In addition, in a 50 kWth chemical looping circulating fluidized bed combustor unit built up by NETL, each stage of the fuel reactor has its own gas chamber, and only the particles can move between two stages, which allows the adjustment of the gas flow introduced to each stage according to the state of the oxygen carriers [12].

2.3   Chemical Looping Combustion with Reverse Flow

The fluidized bed reactor has been widely used for the CLC process in view of the advantages of stable solid circulation, high gas-solid heat and mass transfer, and favorable operation performance. Most of the existing research for CLC unit with fluidized beds was conducted in the laboratory-scale units and at atmospheric pressure. Further study will be carried out to design and operate systems which can realize continuous solid circulation and flexible adjustment at the industrial scale conceptualized for power generation.

Compared with interconnected fluidized beds, the fixed bed configuration is a much simpler design for CLC. The oxygen carrier is statically loaded in the fixed bed, while oxidizing and reducing gases are introduced to the reactor alternatively. High-pressure operation can be easily realized in the fixed bed because solid circulation is not needed. The fixed bed reactor is more compact than the fluidized bed, and particles are not fluidized, which is beneficial for better utilization of the oxygen carrier, lower capital cost, and smaller process footprint. In addition, the issue of attrition and gas-solid separation is avoided, and the CLC process can be easily applied to existing bench-scale units without considerably complicated equipment. However, it is difficult to control the dynamic operation performance, and special valves should be used to resist high temperature, which is the main disadvantage of the fixed bed design.

In the fixed bed reactor, the successive cycles of reduction, oxidation, and heat removal are operated. In the reduction process, the gaseous fuel is introduced to react with the oxidized particles. The products are mainly reduced oxygen carrier, CO_2, and H_2O. The type of fuel and oxygen carrier determines if the reduction process is endothermic or exothermic. The reduction process will be stopped when the fuel conversion or CO_2 selectivity reaches a critical value. Inert gas is then fed to the reactor to take away all the flammable gases. In the oxidation process, the air is introduced to commence the oxidation process to realize the regeneration of reduced oxygen carriers. As the oxidation process is exothermic, heat removal is needed to avoid the high temperature in the reactor. Air is continuously blowing through the reactor until most of heat is removed. Figure 2.5 displays the simplified diagram of reverse-flow operation in a fixed bed CLC reactor. The gases flow via the path along "1-2-R-3-4" is the normal flow condition, and the gases will flow through "1-2'-R-3'-4" when the direction is switched. The reverse-flow operation mode is beneficial for improving the power generation efficiency of a downstream gas turbine cycle.

To enhance the reaction efficiency, a fixed bed reactor can be split into small reactor modules emulating the performance of a simulated moving bed (SMB) reactor. Figure 2.6 displays a simplified diagram of a CLC reactor train operating in simulated moving bed (SMB) mode. A number of fixed bed modules are included. Valves are located in the inlets and outlets to connect all the modules. In the reduction process, the fuel valve is open, and the other two valves are closed. Air

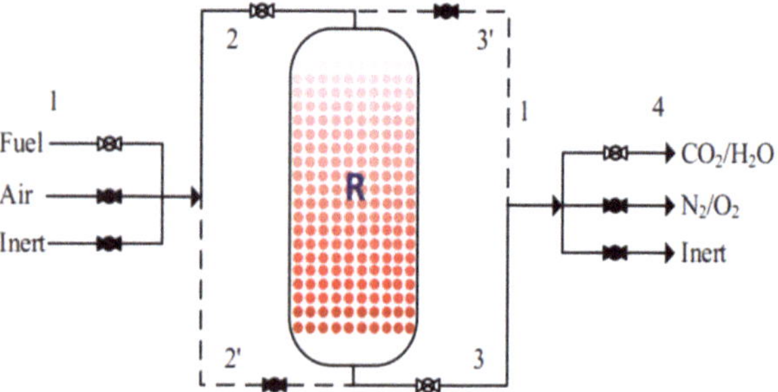

Fig. 2.5 Simplified diagram of reverse-flow operation in a fixed bed CLC reactor. Forward flow pass is 12-R-3-4 and backward flow pass is 1–2'-R-3'-4 [13]

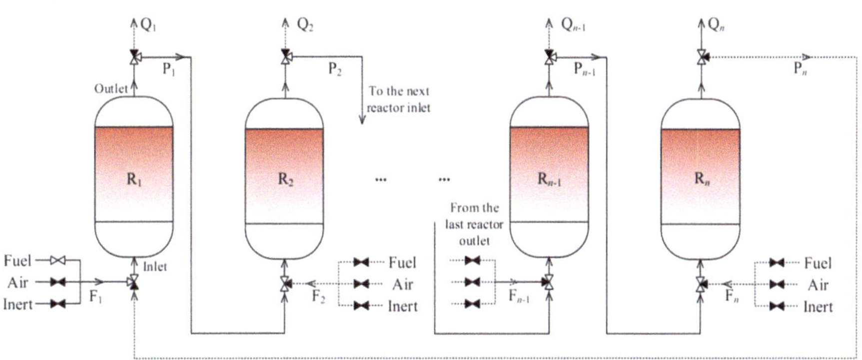

Fig. 2.6 Simplified diagram of a CLC reactor train operating in a SMB mode [14]

is introduced to the first reactor R_1 and flows through the path of $F_1 \rightarrow R_1 \rightarrow P_1 \rightarrow R_2 \rightarrow P_2 \rightarrow \cdots \rightarrow R_{n-1} \rightarrow P_{n-1} \rightarrow R_n \rightarrow Q_n$. Once the reduction stage in R_1 is completed, the fuel is sent to the second reactor R_2, and the gas flows through $F_2 \rightarrow R_2 \rightarrow P_2 \rightarrow \cdots \rightarrow R_{n-1} \rightarrow P_{n-1} \rightarrow R_n \rightarrow P_n \rightarrow R_1 \rightarrow Q_1$ by switching inlet and outlet port valves. The oxidation process will be started until the reduction of oxygen carrier is completed in the last reactor R_n. The similar control of valves is conducted by introducing the air through the air valve. In the heat removal and purge stages, air flows through $F_1 \rightarrow R_1 \rightarrow P_1 \rightarrow R_2 \rightarrow P_2 \rightarrow \cdots \rightarrow R_{n-1} \rightarrow P_{n-1} \rightarrow R_n \rightarrow Q_n$, pushing the residual heat or gases out of each reactor module. The reverse-flow operation can also be achieved in the SMB reactor with the gas introduced to the last reactor R_n first and products exiting from the outlet of the first reactor R_1. The improvement of energy efficiency can be achieved while satisfying constraints of carbon capture, feed conversion, and exit gas temperature variations. In addition, the profiles of temperature and heat conversion are more uniform during the oxidation and heat removal process under the reverse-flow operating condition.

2.4 Oxygen Carriers for CLC

In a chemical looping combustion system, the oxygen carrier plays the role of transferring heat and oxygen from the air reactor to the fuel reactor. The oxygen carrier is often porous, which has low transfer resistance not only within the intraparticle pores but also between the particle surface and external atmosphere. An oxygen carrier should be able to generate oxygen ions or vacancies and electrons or holes. In the reduction process, the oxygen carrier gets contact with fuel, and the oxygen anions in the surface react with the fuels to produce CO_2 and H_2O. Thereby, an oxygen chemical potential gradient is formed between the surface and bulk of the oxygen carrier, causing more O^{2-} to migrate to the surface. At the same time, electrons move to the center of the particle to maintain the overall charge neutrality.

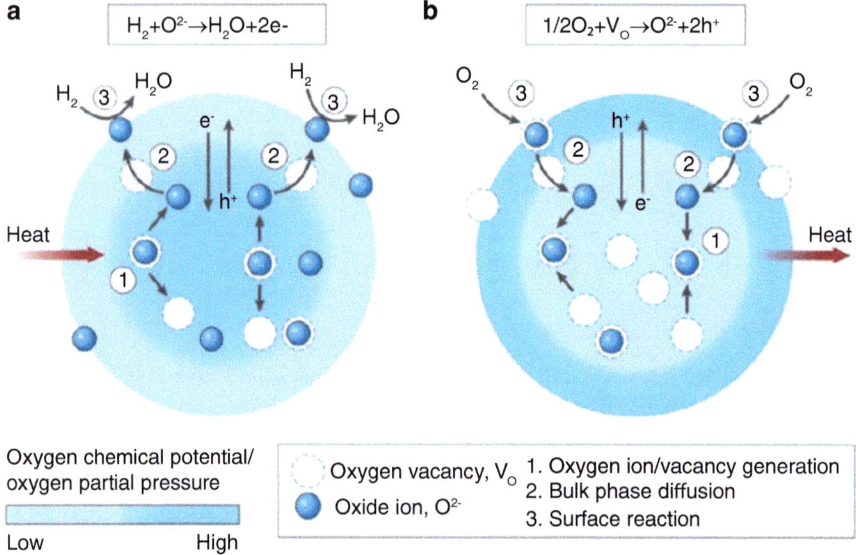

Fig. 2.7 Oxygen transfer mechanism of oxygen carrier during (**a**) reducing and (**b**) oxidizing process [15]

During the oxidation process, the electrons migrate from the bulk to the surface. The O^{2-} is generated to refill the oxygen vacancy, so the oxidization of the oxygen carrier is realized (Fig. 2.7).

For a high-efficiency CLC system, it is of vital importance to select a type of appropriate oxygen carrier. The most important property of an oxygen carrier is to have good reactivity in both of the oxidation and reduction processes. Besides, the oxygen carrier used for CLC should also have the following properties [16]:

1. Be stable after redox cycle under high temperature.
2. Have good mechanical resistance to crushing and abrasion in continuous circulation in fluidized beds.
3. Have good resistance to sintering, agglomeration, and carbon deposition.
4. Be abundant and economically feasible.
5. Be environmentally friendly during preparation and subsequent processing in order not to cause secondary pollution to the environment.

The commonly used oxygen carriers can be categorized into six types according to the active components including the nickel-based, copper-based, iron-based, manganese-based, cobalt-based, and calcium-based oxygen carriers. Among these oxygen carriers, the main active component of calcium-based oxygen carriers is $CaSO_4$, while the active components of other five oxygen carriers are their corresponding metal oxides. The oxygen carriers have different states in the oxidation-reduction reaction, and their oxygen-carrying capacities are also different.

Table 2.1 Oxygen-carrying capacity for different metal/metal oxide pairs

Oxygen carrier	Oxygen transport capacity R_O
NiO/Ni	0.21
CuO/Cu_2O	0.10
CuO/Cu	0.20
Fe_2O_3/Fe_3O_4	0.03
Fe_2O_3/FeO	0.10
Fe_2O_3/Fe	0.30
Mn_3O_4/MnO	0.07
Co_3O_4/Co	0.27
CoO/Co	0.21
$CaSO_4/CaS$	0.47

The oxygen-carrying capacity R_O is defined as the mass ratio of lattice oxygen that can be utilized in the oxygen carrier for CLC. R_O is expressed as $R_O = (m_{ox} - m_{re})/m_{ox}$, where m_{ox} and m_{re} represent the mass of the oxygen carrier in the fully oxidized state and reduced state, respectively. The oxygen-carrying capacities R_O of several typical oxygen carriers are given in Table 2.1.

2.4.1   Metal Oxide Oxygen Carriers

Each type of oxygen carrier has its own advantages and disadvantages. There hasn't been a type of ideal oxygen carrier that possesses all the favorable properties. The melting points of the common oxygen carriers are in the order of Ni-based>Co-based>Mn-based>Fe-based>Cu-based. The reactivity of oxygen carriers is in the order of Ni-based>Cu-based>Fe-based>Mn-based. The operating temperature in the CLC reactor is about 700–1100 °C. Generally, the pure metal oxide cannot withstand such high temperature, and the low-melting material will easily lead to slagging and sintering, which makes the specific surface area and the reaction rate decrease rapidly. In order to obtain high-performance oxygen carrier, the synthetic oxygen carrier is usually prepared by mixing active metal oxides with inert materials such as alumina, magnesium aluminate, zirconia, bentonite, titania, silica, etc. The active components are dispersed on the inert material. The oxygen carrier with the composite structure has the advantages of larger active surface and higher mechanical strength to resist friction compared to the pure metal oxides.

The Fe-based oxygen carrier is environmentally friendly, nontoxic, and inexpensive, which has attracted extensive attention. In addition, the reactivity of Fe-based oxygen carrier is stable under different conditions. The Fe-based oxygen carrier can be reduced to different states, and the conversion from Fe_2O_3 to Fe_3O_4 is faster than conversion to FeO or Fe. Therefore, the conversion from Fe_2O_3 to Fe_3O_4 is usually used in the CLC process. Compared to other oxygen carriers, Fe-based oxygen carrier of Fe_2O_3/Fe_3O_4 has smallest oxygen-carrying capacity which is about 0.03.

Because of the weak redox reactivity and low oxygen-carrying capacity, the complete conversion of fuel is difficult to be achieved, and large amount of oxygen carrier is needed to transfer sufficient lattice oxygen to the fuel. Besides, the slagging problem is another bottleneck of the application of Fe-based oxygen carrier for large-scale CLC unit.

The reactivity of Ni-based oxygen carrier is high, especially reacting with methane. The high reactivity can be maintained at high temperature of 900–1000 °C. Metal Ni is a catalyst for methane reforming reaction, which may be the main reason for its high reactivity to methane. Ni-based oxygen carrier also shows good thermal stability. However, the price of Ni-based oxygen carrier is high and Ni has strong biological toxicity.

The Cu-based oxygen carrier has high oxygen-carrying capacity and high reactivity. The reduced state of Cu-based oxygen carrier is Cu_2O and Cu. Addition heat is not needed to supply to the system using Cu-based oxygen carrier, because both of the oxidation and reduction processes are exothermic. CuO has the property of releasing molecular oxygen at high temperature, which can be applied for the CLOU process. Cu-based oxygen is cost-effective and has little environmental pollution. However, the oxidation rate is low for pure CuO oxygen carrier, and it will decrease drastically after several redox cycles because of the low-melting point and sintering at high temperature. Even with the support of some inert materials, the Cu-based oxygen carrier is still prone to degradation in continuous redox cycles, which is the core issue restricting its further application.

Mn-based oxygen carrier is inexpensive and nontoxic, and it is also regarded as a promising oxygen carrier used for CLC. Similar as Fe-based oxygen carrier, Mn-based oxygen carrier is limited by its thermodynamic property and low oxygen-carrying capacity. The reactivity of Mn-based oxygen carrier is poor especially when reacting with coal and methane.

Co-based oxygen carrier has high oxygen-carrying capacity, but it is toxic and expensive. Little research can be found in the existing literature.

2.4.2 Other Oxygen Carriers

$CaSO_4$ has its unique advantage as an oxygen carrier in terms of oxygen-carrying capacity of 0.47, which is much higher than the metal oxide oxygen carriers. In the CLC process, the main reactions are as follows:

$$\frac{(2n+m)}{4} CaSO_4 + C_nH_{2m} \rightarrow \frac{(2n+m)}{4} CaS + mH_2O + nCO_2 \qquad (2.9)$$

$$\frac{(2n+m)}{4} CaS + \left(n + \frac{m}{2}\right)O_2 \rightarrow \frac{(2n+m)}{4} CaSO_4 \qquad (2.10)$$

However, the reactivity and chemical stability of Ca-based oxygen carriers are not good. The loss of S in the redox cycles will reduce the reactivity and oxygen-carrying capacity. SO_2 and other pollutant gases will be produced during the oxidation process.

Since the single metal oxide oxygen carrier has some defects in the performance, the mixed-metal oxides oxygen carrier formed by doping two or more metal oxides is also used. The mixed-metal oxygen carrier shows better performance in terms of larger specific surface area, stronger mechanical strength, and larger oxygen-carrying capacity. For example, to improve the reactivity of Fe-based oxygen carrier, different supports can be used including alkali metal materials (Li, Na, K), alkaline earth metal materials (Mg, Ca), transition metal materials (Sc, Ti, V, Cr, Mn, Co, Ni, Cu, Zn, Y, Zr), and rare earth metal materials (La, Ce).

2.5   Calcium Looping Combustion

Besides chemical looping, another second-generation carbon capture technology is the calcium looping technology which was proposed by Shimizu et al. [17] in 1999. The calcium oxide sorbents used in calcium looping process are often from the decomposition of limestone or dolomite which is low-priced and globally available [18].

In the calcium looping process, two reaction steps, namely, carbonation and calcination, are included in the carbonator and calciner, respectively. Carbonation is calcination in reverse. In the carbonation process, CaO reacts with the CO_2 from the flue gas to form $CaCO_3$ in the temperature between 580 °C and 700 °C, while in the calcination process, the solid calcium carbonate thermally decomposes into CO_2 and CaO in the temperature between 850 °C and 950 °C, as displayed in Eqs. 2.11 and 2.12 [18]. The high-concentration CO_2 can be obtained for subsequent transport and the final storage and/or utilization, while the regenerated sorbent is recycled in the calcium looping process. The deactivated calcium-based sorbent after the calcium looping cycle can be used in the cement industry (Fig. 2.8):

$$CaO + CO_2 \rightarrow CaCO_3 \; \Delta H = -178 \, kJ/mol \tag{2.11}$$

$$CaCO_3 \rightarrow CaO + CO_2 \; \Delta H = 178 \, kJ/mol \tag{2.12}$$

The post-combustion CO_2 capture system based on calcium looping technology has some obvious advantages over solvent-based CO_2 capture technologies. In the calcium looping technology, the reaction rate is faster due to high operating temperature, and the CO_2 adsorption capacity and carbon capture efficiency are high. The $CaCO_3$ and CaO are less hazardous to health when compared with solvents. Since it is not necessary to pressurize or cool the flue gas, the calcium looping process can be directly applied to the industrial power plants. In addition, the mature

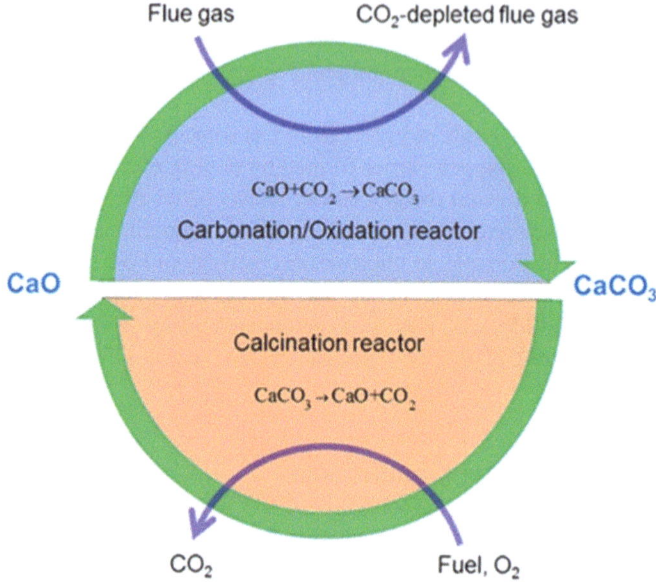

Fig. 2.8 Schematic diagram of calcium looping pocess [19]

circulating fluidized bed technology can be used for calcium looping process, such as the 1.0 MWth unit in Darmstadt, Germany [20]; the 1.7 MWth unit in La Pereda, Spain [21]; and the 1.9 MWth unit at the Industrial Technology Research Institute of Taiwan, China [22]. This has commercially proved the feasibility and reliability of the circulating fluidized bed technology for calcium looping process and saved considerable technical research costs.

However, the calcium looping technology also has some disadvantages. The main drawback is the CaO sorbent regenerability over multiple high-temperature cycles [18]. After a number of calcination-carbonation cycles, the structure of the sorbents will be gradually changed, which results in the decrease of the sorbent reactivity and carbonation conversion. In addition, the porosity of the sorbent will decrease because of sintering. The micropores that are beneficial to absorb CO_2 decreases, while the mesoporous increases. Furthermore, the particle distribution of the sorbents will also be changed, because the particles are subjected to attrition/fragmentation phenomena caused from the thermal stress and mechanical stress in a fluidized bed. The fine particles are easily to be carried out of the reactor, resulting in the loss of the sorbents.

In recent years, the magnesium looping process analogous to the calcium looping process in which MgO is employed instead of CaO has also been proposed. It is not discussed here but the reader is referred to the literature [23].

References

1. H. Kruggel-Emden, F. Stepanek, A. Munjiza, A study on the role of reaction modeling in multi-phase CFD-based simulations of chemical looping combustion. Oil Gas Sci. Technol. d'IFP Energies Nouv. **66**(2), 313–331 (2011)
2. J. Adánez, A. Abad, T. Mendiara, P. Gayán, L.F. de Diego, F. García-Labiano, Chemical looping combustion of solid fuels. Prog. Energy Combust. Sci. **65**, 6–66 (2018)
3. T. Mattisson, A. Lyngfelt, H. Leion, Chemical-looping with oxygen uncoupling for combustion of solid fuels. Int. J. Greenh. Gas Control **3**(1), 11–19 (2009)
4. I. Adánez-Rubio, A. Abad, P. Gayán, L.F. De Diego, F. García-Labiano, J. Adánez, Identification of operational regions in the chemical-looping with oxygen uncoupling (CLOU) process with a Cu-based oxygen carrier. Fuel **102**, 634–645 (2012)
5. M. Rydén, H. Leion, T. Mattisson, A. Lyngfelt, Combined oxides as oxygen-carrier material for chemical-looping with oxygen uncoupling. Appl. Energy **113**, 1924–1932 (2014)
6. P. Gayán, I. Adánez-Rubio, A. Abad, L.F. De Diego, F. García-Labiano, J. Adánez, Development of Cu-based oxygen carriers for chemical-looping with oxygen uncoupling (CLOU) process. Fuel **96**, 226–238 (2012)
7. T. Song, L. Shen, Review of reactor for chemical looping combustion of solid fuels. Int. J. Greenh. Gas Control **76**, 92–110 (2018)
8. T. Song, L. Shen, Review of reactor for chemical looping combustion of solid fuels. Int. J. Greenh. Gas Control **76**(April), 92–110 (2018)
9. H.R. Kim, D. Wang, L. Zeng, S. Bayham, A. Tong, E. Chung, M.V. Kathe, S. Luo, O. McGiveron, A. Wang, et al., Coal direct chemical looping combustion process: design and operation of a 25-KW$_{Th}$ sub-pilot unit. Fuel **108**, 370–384 (2013)
10. A. Thon, M. Kramp, E.U. Hartge, S. Heinrich, J. Werther, Operational experience with a system of coupled fluidized beds for chemical looping combustion of solid fuels using ilmenite as oxygen carrier. Appl. Energy **118**, 309–317 (2014)
11. H. Gu, L. Shen, S. Zhang, M. Niu, R. Sun, S. Jiang, Enhanced fuel conversion by staging oxidization in a continuous chemical looping reactor based on iron ore oxygen carrier. Chem. Eng. J. **334**, 829–836 (2018)
12. R. Siriwardane, J. Riley, S. Bayham, D. Straub, H. Tian, J. Weber, G. Richards, 50-KWth methane/air chemical looping combustion tests with commercially prepared CuO-Fe2O3-alumina oxygen carrier with two different techniques. Appl. Energy **213**, 92–99 (2018)
13. L. Han, G.M. Bollas, Chemical-looping combustion in a reverse-flow fixed bed reactor. Energy **102**, 669–681 (2016)
14. C. Chen, G.M. Bollas, Design and scheduling of semibatch chemical-looping reactors. Ind. Eng. Chem. Res. **59**(15), 6994–7006 (2020)
15. L. Zeng, Z. Cheng, J.A. Fan, L.-S. Fan, J. Gong, Metal oxide redox chemistry for chemical looping processes. Nat. Rev. Chem. **2**(11), 349–364 (2018)
16. M.M. Hossain, H.I. de Lasa, Chemical-looping combustion (CLC) for inherent CO2 separations—a review. Chem. Eng. Sci. **63**(18), 4433–4451 (2008)
17. T. Shimizu, T. Hirama, H. Hosoda, K. Kitano, M. Inagaki, K. Tejima, A twin fluid-bed reactor for removal of CO2 from combustion processes. Chem. Eng. Res. Des. **77**(1), 62–68 (1999)
18. D.P. Hanak, E.J. Anthony, V. Manovic, A review of developments in pilot-plant testing and modelling of calcium looping process for CO2 capture from power generation systems. Energy Environ. Sci. **8**, 2199 (2015)
19. J. Chen, L. Duan, Z. Sun, Review on the development of sorbents for calcium looping. Energy Fuel **34**(7), 7806–7836 (2020)

20. J. Ströhle, M. Junk, J. Kremer, A. Galloy, B. Epple, Carbonate looping experiments in a 1 MWth pilot plant and model validation. Fuel **127**, 13–22 (2014)
21. B. Arias, M.E. Diego, J.C. Abanades, M. Lorenzo, L. Diaz, D. Martínez, J. Alvarez, A. Sánchez-Biezma, Demonstration of steady state CO2 capture in a 1.7 MWth calcium looping pilot. Int. J. Greenh. Gas Control **18**, 237–245 (2013)
22. M.H. Chang, W.C. Chen, C.M. Huang, W.H. Liu, Y.C. Chou, W.C. Chang, W. Chen, J.Y. Cheng, K.E. Huang, H.W. Hsu, Design and experimental testing of a 1.9 MWth calcium looping pilot plant. Energy Procedia **63**, 2100–2108 (2014)
23. K. Rausis, A.R. Stubbs, I.M. Power, C. Paulo, Rates of atmospheric CO_2 capture using magnesium oxide powder. Int. J. Greenh. Gas Control **119**, 103701 (2022)

Chapter 3
Process Simulations and Techno-Economic Analysis with Aspen Plus

3.1 Aspen Plus

The Aspen (Advanced System for Process Engineering) Plus is a process simulation software [1] which is used for the design, development, analysis, and optimization of technical processes in chemical plants, environmental systems, power stations, complex manufacturing operations, chemical and biological processes, and similar technical functions.

Process simulation is a model-based representation of chemical, physical, biological, and other technical processes and unit operations in the Aspen Plus software. Basic prerequisites for the model are chemical and physical properties of pure components and mixtures, of reactions, and of mathematical models which, in combination, allow the calculation of process properties by the software. The software describes processes in flow diagrams where unit operations are positioned and connected by product or educt streams. The software solves the mass and energy balance to find a stable operating point on specified parameters. The goal of a process simulation is to find the optimal conditions for a process. This is essentially an optimization problem which is solved by an iterative process.

The applications of the Aspen Plus software are described in this chapter, which uses basic engineering relationships such as mass and energy balance and multiphase and chemical reaction models in modeling a process such as chemical or calcium looping combustion at system level. The effects of different types of oxygen carriers and fuels (gas/solid/biomass) on the performance of various looping systems are discussed under different operating conditions.

In case of the chemical looping combustion process, the char gasification step required when using the solid fuels such as coal is slow and often incomplete, which limits the rate of fuel conversion. The concept of multi-staged fuel reaction has been proposed as an improvement to the original single-stage chemical looping combustion concept to address this issue. The results of the simulations have shown that the multi-staging allows the use of multiple smaller reactors with the same total volume

R. K. Agarwal, Y. Shao, *Modeling and Simulation of Fluidized Bed Reactors for Chemical Looping Combustion*, https://doi.org/10.1007/978-3-031-11335-2_3

without incurring any penalty on the net energy output. In addition, techno-economic analysis of the multi-staged fuel reactor system is described which shows that the two-stage system is more economically viable. For the calcium looping process, process models are described to determine and compare the energy penalty in two types of systems including post-combustion capture and pre-combustion capture. The techno-economic analysis is also presented.

3.2 iG-CLC Process Simulation

3.2.1 Validation of the iG-CLC Process Simulation

The validation of the process simulation of the in situ gasification chemical looping combustion (iG-CLC) process in Aspen Plus against the experimental work of Sahir et al. [2] is described in this section. Since coal is designated as a nonconventional solid in Aspen Plus, its attributes are defined based on the physical and chemical properties of the coal used. The RYIELD reactor block is employed in Aspen Plus to decompose the nonconventional material–coal—into its constituent conventional materials to be able to simulate their reactions. Mass percentages for the component yields are defined based on the proximate and ultimate analysis of the bituminous Colombian coal used in the experiment of Sahir et al. [2] and is summarized in Table 3.1.

The schematic of the flow sheet for this simulation is shown in Fig. 3.1. The coal is first pulverized and dried before it is pressurized and introduced into a shell gasifier to be partially oxidized to form syngas. The oxygen carrier material used is a mixture of 60% Fe_2O_3 by mass and 40% Al_2O_3 by mass as support. The

Table 3.1 Physical and chemical properties of Colombian coal

Proximate analysis (wt. %)				Ultimate analysis (wt. %)					Energy
Moisture	Volatile matter	Fixed carbon	Ash	C	H	N	S	O	LHV (MJ/kg)
3.3	37.0	54.5	5.2	80.7	5.5	1.7	0.6	11.5	29.1

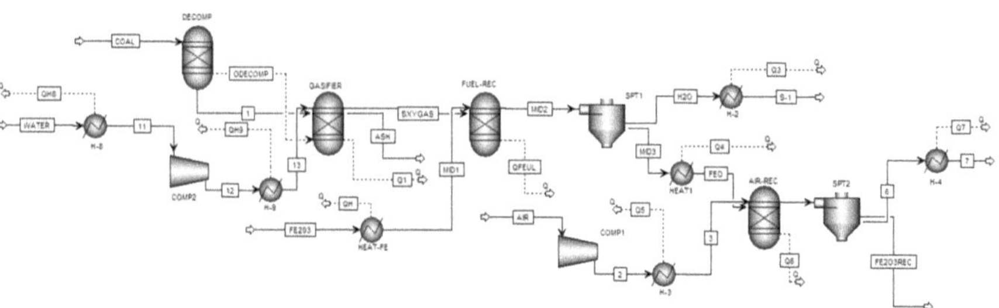

Fig. 3.1 Overall flow sheet of iG-CLC process in Aspen Plus

Table 3.2 Process models used in different parts of the iG-CLC process in Aspen Plus

Name	Model	Function	Reaction formula
DECOMP	RYIELD	Coal devolatilization	Coal → volatile matter + char
BURN	RGIBBS	Gasification	Char + volatile matter + H_2O → CO + H_2
FUEL-R	RGIBBS	Syngas combustion	$3Fe_2O_3 + CO → 2Fe_3O_4 + CO_2$, $3Fe_2O_3 + H_2 → 2Fe_3O_4 + H_2O$
AIR-R	RGIBBS	Carrier re-oxidation	$4Fe_3O_4 + O_2 → 6Fe_2O_3$

supporting material is inert and serves only to increase the reactivity and durability of the oxygen carrier. The molar ratio of steam and carbon is maintained at unity for the process model. The syngas composition at the gasifier outlet is 34.5% CO, 50.3% H_2, 12.3% H_2O, and 2.4% CO_2. The syngas is converted completely to CO_2 and H_2O in the fuel reactor, while the Fe_2O_3 in the oxygen carrier is reduced to Fe_3O_4. The outflow from the fuel reactor is a concentrated stream of H_2O and CO_2. After condensing the stream, high-purity CO_2 is obtained. The reduced oxygen carrier is fed into the air reactor where the oxidation reaction takes place with an 80% conversion of Fe_3O_4 to Fe_2O_3.

The various process models used in the Aspen Plus flow sheet, shown in Fig. 3.1, are summarized in Table 3.2. The coal devolatilization is defined by the RYIELD reactor, followed by the gasification of coal represented by the RGIBBS reactor. Another RGIBBS reactor defines the actual syngas combustion and the corresponding reduction of the oxygen carrier. These blocks together represent the fuel reactor. The flow sheet within the Aspen Plus simulation package cannot model this entire reaction with one reactor. As a result, the fuel reactor simulation is broken down into several different reactor simulations. The air reactor is also modeled as an RGIBBS reactor.

The energy balance of the iG-CLC process model was analyzed using the input values from the experiment of Sahir et al. [2] The input values and the energy requirements for the various units and streams in Fig. 3.1 are presented in Table 3.3. Energy is consumed mainly in the compressor processes. Compressed air is required in the air reactor to regenerate Fe_2O_3 from Fe_3O_4; the air compressor for the combustor compresses air to 18 atm. Another compressor is used to compress the steam for the gasifier. There is a large amount of energy produced in the air reactor, but the fuel reactor needs to be supplied with energy. This is because the net heat work in the fuel reactor is the summation of the heat work from the DECOMP, GASIFER, and FUEL-R blocks in Aspen flow sheet. Although FUEL-R produces energy because of the combustion of syngas, the combined energy requirement of DECOMP and GASIFIER is more than the energy produced in FUEL-R. Summing the energy requirements of each individual stream, the total energy obtained from the iG-CLC process is 554.2 kW.

The results shown in Table 3.3 for the baseline case with a coal feed rate of 100 kg/h are in excellent agreement with those reported by Sahir et al. [2] Hence, these calculations validate the iG-CLC process simulation developed in Aspen Plus.

Table 3.3 Input and output values for the baseline case corresponding to the work of Sahir et al. [2]

Input values	Coal (kg/h)	100
	Steam (kg/h)	140
	Airflow rate (kg/h)	71
	Temperature of fuel reactor (°C)	950
	Temperature of air reactor (°C)	935
	Fe_2O_3 flow in the fuel reactor (kg/h)	5921
	Al_2O_3 in the system (kg/h)	3951
	Particle density (kg/m^3)	3200
Energy balance (kW)	Fuel reactor	-161.8
	Air reactor	688.0
	Cool air reactor exhaust	135.4
	Cool flue gas	148.3
	Cool oxygen carrier for air reactor	40.9
	Reheat oxygen carrier for fuel reactor	-42.7
	Heat steam	-69.8
	Heat air	-184.1
	Net	554.2

Table 3.4 Physical and chemical properties of three ranks of coal

	Proximate analysis (wt. %)				Ultimate analysis (wt. %)					Energy
Coal rank	Moisture	Volatile matter	Fixed carbon	Ash	C	H	N	S	O	LHV (MJ/kg)
Anthracite	1.0	7.5	59.9	31.6	60.7	2.1	0.9	1.3	2.4	21.9
Bituminous	2.3	33.0	55.9	8.8	65.8	3.3	1.6	0.6	17.6	21.9
Lignite	12.6	28.6	33.6	25.2	45.4	2.5	0.6	5.2	8.5	16.3

3.2.2 Energy Output of Different Ranks of Coals

It is of interest to investigate the performance of the iG-CLC plant for different ranks of coal. The three coal ranks considered in this section, in the order of carbon content from most to least, are anthracite, bituminous, and lignite. The proximate analysis and ultimate analysis of these coals are summarized in Table 3.4.

For each coal rank in Table 3.4, the coal feeding rate is held constant at 100 kg/h, while the airflow rate is varied in order to evaluate the effect of air supply on the energy output. The results are presented in Fig. 3.2. For each coal rank, an increase in the airflow rate leads to an increase in the net energy output of the iG-CLC plant until a maximum is achieved. If the airflow rate is increased further, the energy output starts to decrease, albeit very slowly. The different ranks of coal each have a different optimum value that corresponds to a maximum energy output on the *y*-axis for a certain airflow rate shown on the *x*-axis. The difference is expected given the different physical and chemical properties of the coals. By qualitative analysis, one can infer that higher concentration of fixed carbon in a coal provides more fuel for combustion and higher concentration of volatile matter and ash requires less energy to decompose the coal.

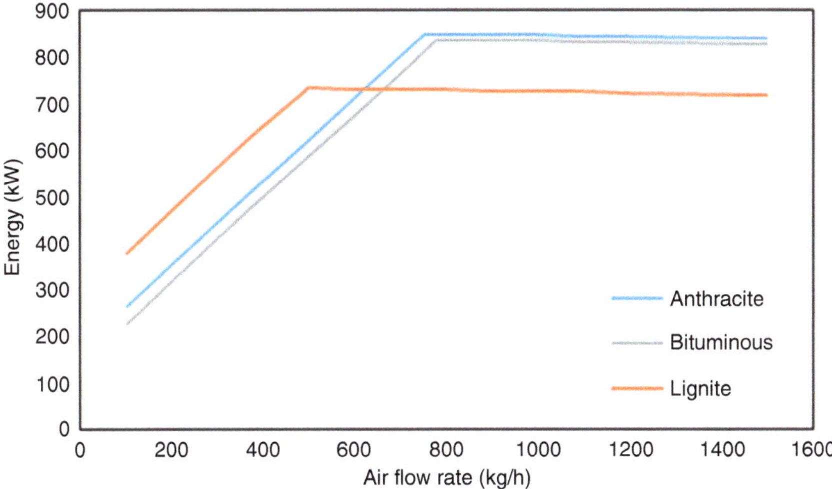

Fig. 3.2 Overall energy output with increasing airflow rate for 100 kg/h for three ranks of coal

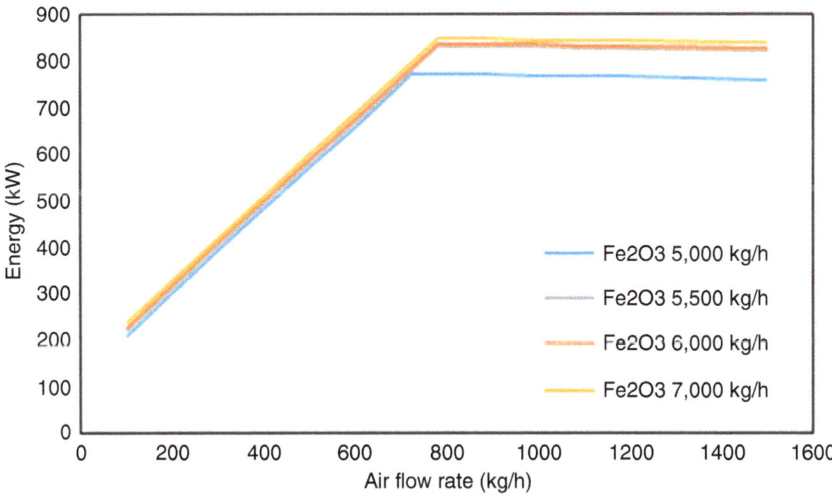

Fig. 3.3 Overall energy output with increasing airflow rate and oxygen carrier feed rate for 100 kg/ h of bituminous coal

The oxygen carrier plays a vital role in the iG-CLC process: it provides oxygen for combusting the syngas in the fuel reactor, and it gets oxidized by atmospheric air in the air reactor. Both these reactions contribute a large amount of energy to the net output of the iG-CLC plant. Figure 3.3 plots the energy output against airflow rate for different oxygen carrier feeding rates in the system for bituminous coal. As expected, it shows that for a given airflow rate, a higher oxygen carrier feed produces more energy. However, when the oxygen carrier feeding rate increases above a

Table 3.5 Maximum energy output and optimum feed ratio for three ranks of coal

Coal rank	Maximum energy (kW)	Optimum ratio of coal, air, and oxygen carrier
Anthracite	841.258	1:8:52
Bituminous	832.373	1:8:55
Lignite	707.905	1:5:35

certain threshold value, the marginal increase in the energy output becomes extremely small. For bituminous coal with a feeding rate of 100 kg/h, the optimum rates of airflow and oxygen carrier feed are approximately 800 kg/h and 5500 kg/h, respectively. In other words, the optimum ratio of coal, air, and oxygen carrier feed by mass is 1:8:55, corresponding to a net energy output of 832 kW.

Similar analysis is presented for the anthracite and lignite coal. Table 3.5 summarizes the maximum energy output with 100 kg/h coal feed rate and optimum ratio of coal, air, and oxygen carrier for the three coal ranks given in Table 3.4.

The results in Table 3.5 provide a starting point to design and operate an iG-CLC plant to maximize power generation for various ranks of coal. Other parameters that may affect the energy output include the temperature and pressure of the reactors, the effect of particle size on reactivity, etc., which are not discussed here in this chapter.

3.3 CLOU Process Simulation

It has been identified that the char gasification is the bottleneck of the fuel conversion rate in the iG-CLC process [3]. An alternative process known as the chemical looping with oxygen uncoupling (CLOU) has been proposed to overcome the low reactivity of the char gasification stage in the coal-direct chemical looping combustion [4–7]. The CLOU process is based on using materials that can release gaseous oxygen in the fuel reactor as the oxygen carriers. These materials can also be regenerated in high-temperature environment, for example, in the air reactor. The combustibles (volatiles and char) produced as a result of coal decomposition are burned directly with the gaseous oxygen released by the oxygen carrier. Thus, in CLOU, the slow gasification step in the chemical looping combustion with solid fuels is avoided, resulting in a much faster solid conversion [6], leading to a reduced solids inventory and circulation rate and increased fuel conversion and CO_2 capture efficiency.

3.3.1 Validation of the CLOU Process Simulation

The validation of process simulation of CLOU using Aspen Plus against the experiment of Abad et al. [8] is described in this section. The solid fuel used in the experiment is a bituminous "El Cerrejon" coal. The coal is subjected to a thermal pre-treatment for pre-oxidation in order to avoid coal swelling and bed

Table 3.6 Properties of bituminous "El Cerrejon" coal

	Proximate analysis (wt. %)				Ultimate analysis (wt. %)						Energy
Components	Moisture	Volatile matter	Fixed carbon	Ash	C	H	N	S	O	Ash	LHV (kJ/kg)
Fresh	7.5	34.0	49.9	8.6	70.8	3.9	1.7	0.5	7.20	15.9	25,880
Pre-treated	2.3	33.0	55.9	8.8	65.8	3.3	1.6	0.6	17.6	11.1	21,899

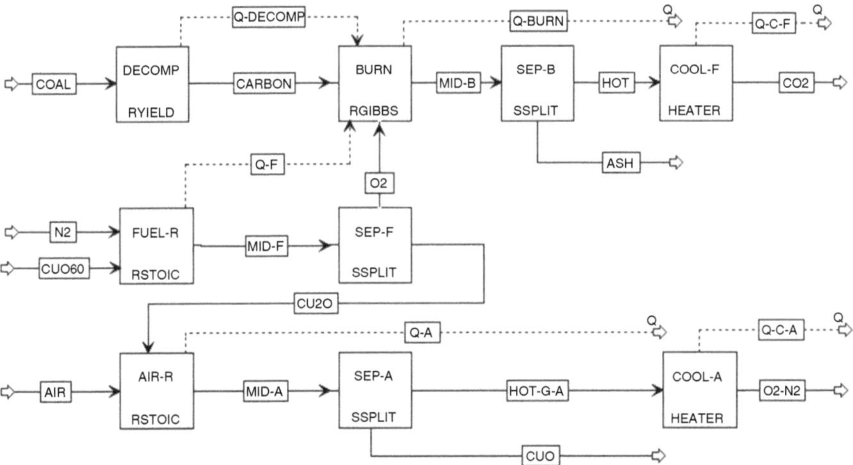

Fig. 3.4 Flow sheet of CLOU process in Aspen Plus for Abad et al.'s experiment [8]

agglomeration. In the experiment, coal was heated at 180 °C in the atmospheric air for 28 h. Proximate and ultimate analyses of the pre-treated coal are given in Table 3.6. Oxygen carrier particles are prepared by spray drying and contain 60 wt. % CuO and 40 wt. % $MgAl_2O_4$ as supporting material.

The Aspen Plus model flow sheet developed for the Abad et al.'s experimental setup [8] is shown in Fig. 3.4.

As given in Table 3.7, coal devolatilization is defined by the RYIELD reactor, followed by the gasification of coal represented by the RGIBBS reactor. The RSTOIC reactor defines the actual fuel combustion. It should be noted here that the three reactor blocks together represent the fuel reactor in Abad et al.'s experiment [8]. The flow sheet within the Aspen Plus simulation package cannot model the entire reaction with one reactor. As a result, the fuel reactor is broken down into several different reactor simulations. The air reactor is modeled as an RSTOIC reactor. The molar flow rate of CuO exiting and Cu_2O feeding in two separate blocks is defined to be identical to represent the circulation of oxygen carrier within the system; such circulation cannot be defined explicitly in the ASPEN Plus modeling. Detailed descriptions of the model setup can be found in the work of Zhou et al. [9].

Table 3.7 Process models used in different parts of CLOU process in Aspen Plus

Name	Model	Function	Reaction formula
DECOMP	RYIELD	Coal devolatilization and gasification	Coal $\rightarrow$ volatile matter + char
BURN	RGIBBS	Syngas and char burn with O_2	Char +volatile matter + $O_2 \rightarrow CO_2 + H_2O$
FUEL-R	RSTOIC	Carrier reduction reaction	$4CuO \rightarrow 2Cu_2O + O_2$
AIR-R	RSTOIC	Carrier oxidation reaction	$2Cu_2O + O_2 \rightarrow 4CuO$
SEP-F	SSPLIT	O_2 and Cu_2O separation	
SEP-A	SSPLIT	CuO and air separation	
SEP-B	SSPLIT	Separation—ash and flue gas	
COOL-F	HEATER	Flue gas cooler—fuel reactor	H_2O (gas) $\rightarrow H_2O$ (liquid)
COOL-A	HEATER	Flue gas cooler—air reactor	

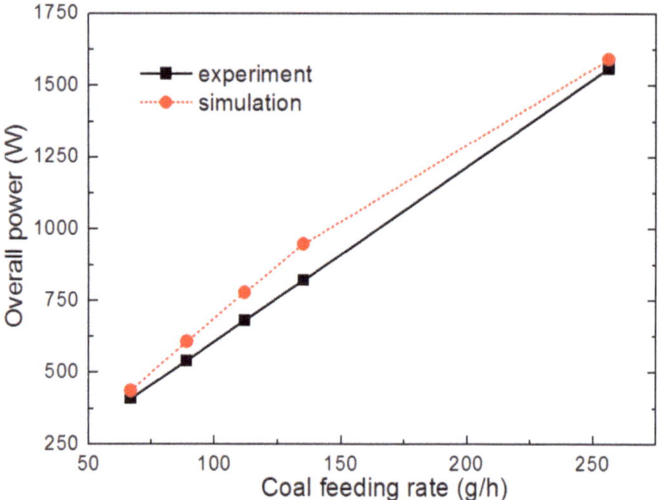

Fig. 3.5 Comparison of overall power between the simulation and the experiment

Figure 3.5 compares the thermal power output from the experiment and the simulation for the entire CLOU system for different coal feeding rates. It can be seen that the estimated power output obtained from the Aspen Plus simulation is in reasonably good agreement with the experimental values. The slight discrepancy between the simulation and the experiment could be attributed to slightly different working conditions in the experimental apparatus that are slightly different from the corresponding block parameters in Aspen Plus. In addition, losses at each location in the experimental apparatus are inevitable, but these cannot be accurately captured in the simulation.

Table 3.8 summarizes the breakdown of power output for various components of the modeled CLOU system in Aspen Plus. Energy is consumed mainly in the compressor processes. Compressed air is required in the air reactor to regenerate

Table 3.8 Thermal analysis at various locations of the modeled CLOU system in Aspen Plus

$\dot{m}_{coal}$ (g/h)	Total power (W)	Q-A (W)	Q-Burn (W)	Q-C-A (W)	Q-C-F (W)	Q-Decomp (W)	Q-F (W)
67	436.6	−175.1	116.4	380	115.3	31.6	−380.1
89	606.4	−79.9	181.9	370.1	134.3	41.7	−477.6
112	777.6	−30.5	296.1	361.1	150.8	53.5	−534.5
135	946.5	51.5	372.7	352.3	170	64.2	−628.8
256	1591.4	180.3	803.6	338.2	269.3	120.7	−1094

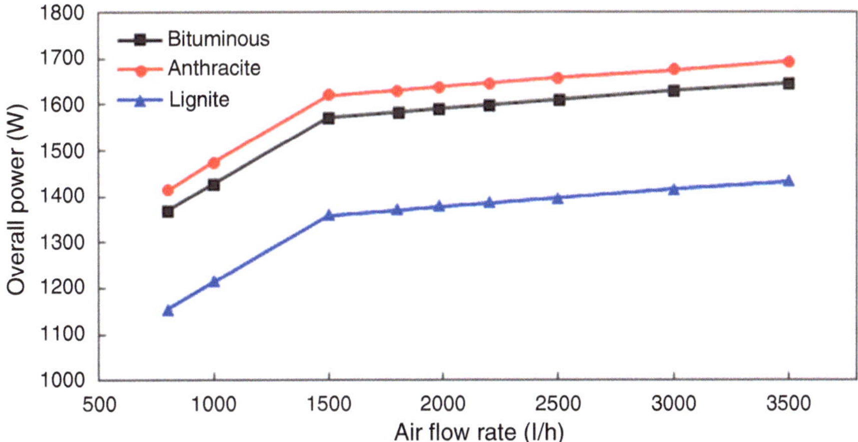

Fig. 3.6 Overall energy output with increasing airflow rate for 256 g/h of three ranks of coal using a Cu-based oxygen carrier

CuO from Cu_2O. Another compressor is used to compress the steam for the gasifier. There is a large amount of energy produced in the air reactor, but the fuel reactor needs to be supplied with energy. This is because the net heat work in the fuel reactor is the summation of the heat work from the DECOMP, BURN, and FUEL-R blocks in Fig. 3.4.

Although BURN produces energy because of the combustion of syngas, the energy requirement of FUEL-R is more than the energy produced in DECOMP and BURN.

3.3.2   Energy Output of Different Ranks of Coal

In order to evaluate the effect of the airflow rate, the coal feeding rate is kept constant at 256 g/h, and the airflow rate is varied. A Cu-based oxygen carrier (CuO/Cu_2O) is used in this model. Figure 3.6 shows the trend in power output with increasing airflow rate for the three different ranks of coal. It can be noted that the power output increases rapidly and linearly with increasing airflow rate until the airflow rate reaches a value of around 1500 l/h, beyond which the increase in power output is very gradual. When the airflow rate is less than 1500 l/h, there is not enough air in the air reactor to re-oxidize the Cu_2O which comes from the fuel reactor. 1500 l/h of air corresponds to the stoichiometric amount of air required to re-oxidize the Cu_2O coming from the fuel reactor completely, which is responsible for releasing the total amount of heat. The reason that the overall power continues to increase slowly for airflow rate greater than 1500 l/h is that the temperature of air reactor is slightly higher than that of the following heat exchanger (which cools the gas out of the air reactor). Therefore, with additional air input, slightly additional energy benefit is obtained. However, it is important to note that in the Aspen Plus model the focus is entirely on heat energy; it does not take into account the mechanical energy consumed by each block of the flow sheet, such as the energy required to supply air into the air reactor. Therefore, the marginal benefit of adding more air into the system beyond the stoichiometric amount of 1500 l/h required to re-oxidize the Cu_2O is likely to be small, and the stoichiometric amount can be considered the optimum airflow rate. The results of Fig. 3.6 are important for estimating the energy output of the CLOU process at the optimum airflow rate for the different ranks of coal. As expected, the coal with the highest carbon content (anthracite) yields the highest energy output. It should also be noted that these results scale linearly for higher coal feeding rates because of the assumptions made in Aspen Plus modeling.

Aside from CuO/Cu_2O, Co- and Mn-based oxygen carriers Co_3O_4/CoO and Mn_2O_3/Mn_3O_4, respectively, are also suitable for CLOU operation since they release gaseous oxygen at high temperatures. It is of interest to investigate the effect of these oxygen carriers on the energy output of the CLOU process. In the case of Co_3O_4 and Mn_2O_3, the oxygen is released according to the following reversible reactions:

$$2Co_3O_4 \rightleftarrows 6CoO + O_2 \tag{3.1}$$

$$6Mn_2O_3 \rightleftarrows 4Mn_3O_4 + O_2 \tag{3.2}$$

For Co_3O_4/CoO and Mn_2O_3/Mn_3O_4, the effect of increasing the airflow is very similar to that of CuO/Cu_2O as shown in Figs. 3.7 and 3.8, respectively. From these figures, it can be seen that the optimum or stoichiometric airflow rate for the Co- and Mn-based oxygen carriers are 1500 and 1800 l/h, respectively.

The maximum power output for each oxygen carrier using the optimum airflow rate and three different ranks of coal is summarized in Table 3.9 considering a constant coal feeding rate of 256 g/h. It should be noted that the feeding rate for each oxygen carrier is varied to ensure that the amount of oxygen released in the fuel reactor remains the same.

For each type of coal, Mn_2O_3 yields the highest energy output, followed by CuO; the energy output of Co_3O_4 is the least. It can be seen that the coal rank has significant impact on overall energy release. The performances of anthracite and bituminous coal in CLOU are similar and higher compared to lignite coal. The

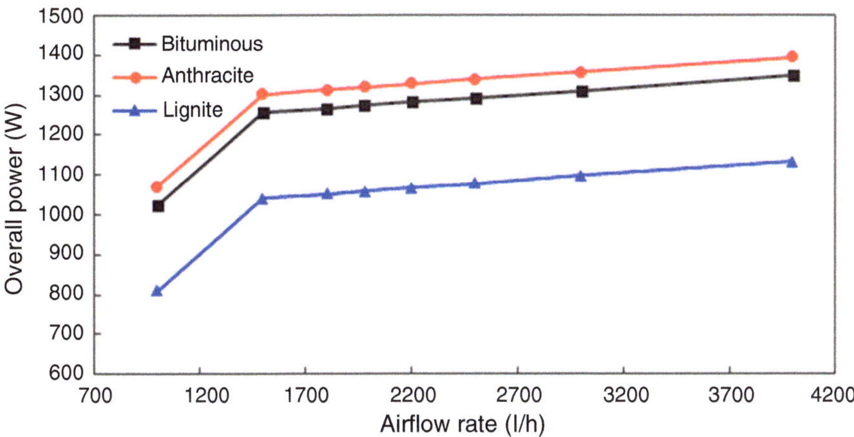

Fig. 3.7 Overall energy output with increasing airflow rate for 256 g/h of three ranks of coal using a Co-based oxygen carrier

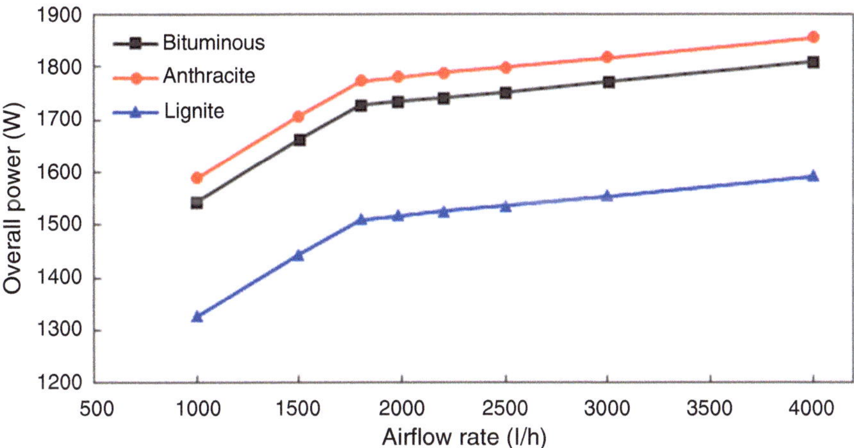

Fig. 3.8 Overall energy output with increasing airflow rate for 256 g/h of three ranks of coal using a Mn-based oxygen carrier

similarity in the performance of anthracite and bituminous coal can be explained by the fact that both have similar carbon content. Char gasification is not a very significant factor in CLOU process performance since the presence of oxygen enables the solid-gas combustion to take place without gasification. The performance of the system can be optimized by utilizing the optimum airflow rate corresponding to the stoichiometric amount of air required to completely re-oxidize the oxygen carrier in the air reactor.

Table 3.9 Maximum energy output based on optimum airflow rate for three ranks of coal and three different oxygen carriers

Coal rank	Oxygen carrier	Optimum airflow rate (l/h)	Oxygen carrier feeding rate (kg/h)	Maximum power output (W)
Anthracite	CuO	1500	9.0	1619.39
	Co_3O_4	1500	13.5	1301.40
	Mn_2O_3	1800	26.0	1773.39
Bituminous	CuO	1500	9.0	1573.71
	Co_3O_4	1500	13.5	1255.75
	Mn_2O_3	1800	26.0	1727.77
Lignite	CuO	1500	9.0	1359.54
	Co_3O_4	1500	13.5	1041.59
	Mn_2O_3	1800	26.0	1510.89

3.4   Multi-staged Fuel Reactor Simulation

CLOU is one solution to avoid the slow char gasification step in the iG-CLC process, which leads to poor coal utilization efficiency since a significant portion of the char remains unreacted and the corresponding energy contained in the fuel is wasted. An alternate solution explored in this section is the option of a multi-staged fuel reaction where the coal is cycled through multiple fuel reactor stages. The multi-staged fuel reaction thus increases the total residence time of the coal in the system resulting in more complete combustion of char and releasing more energy from the fuel. Figure 3.9 shows a conceptual schematic of a two-stage fuel reaction system for CLC. In this design, the oxygen carrier from the air reactor is distributed among the two fuel reactor stages, and the reduced carrier is returned to the air reactor. A four-stage system is very similar to the two-stage system shown in Fig. 3.9, with four fuel reactors instead of two.

In the Aspen Plus flow sheet of the CLC process presented in Sect. 3.1, the gasification step is represented by the RGIBBS reactor block, which performs a complete equilibrium calculation based on the minimization of Gibbs free energy. As a result, the simulation cannot take into account the slow rate of the gasification reaction and instead considers total conversion of char during the gasification process. This section improves the accuracy of the CLC simulation by employing the rate-controlled RPLUG reactor to represent the char gasification step. The coal considered in these simulations is the bituminous Colombian coal used in the work of Sahir et al. [2]. The flow sheet for the revised single-stage CLC model developed in Aspen Plus using the RPLUG reactor is shown in Fig. 3.10.

Table 3.10 summarizes the main reactors used in the Aspen Plus flow sheet and the operating conditions used in the simulations. The coal is introduced to the RYIELD reactor that decomposes the coal into volatile matter and char based on the proximate and ultimate analysis of the coal. The products of the decomposition reactor are mixed with steam at 950 °C and 10 atmospheres in the gasifier. As mentioned previously, this study models the gasifier as RPLUG reactor in order to

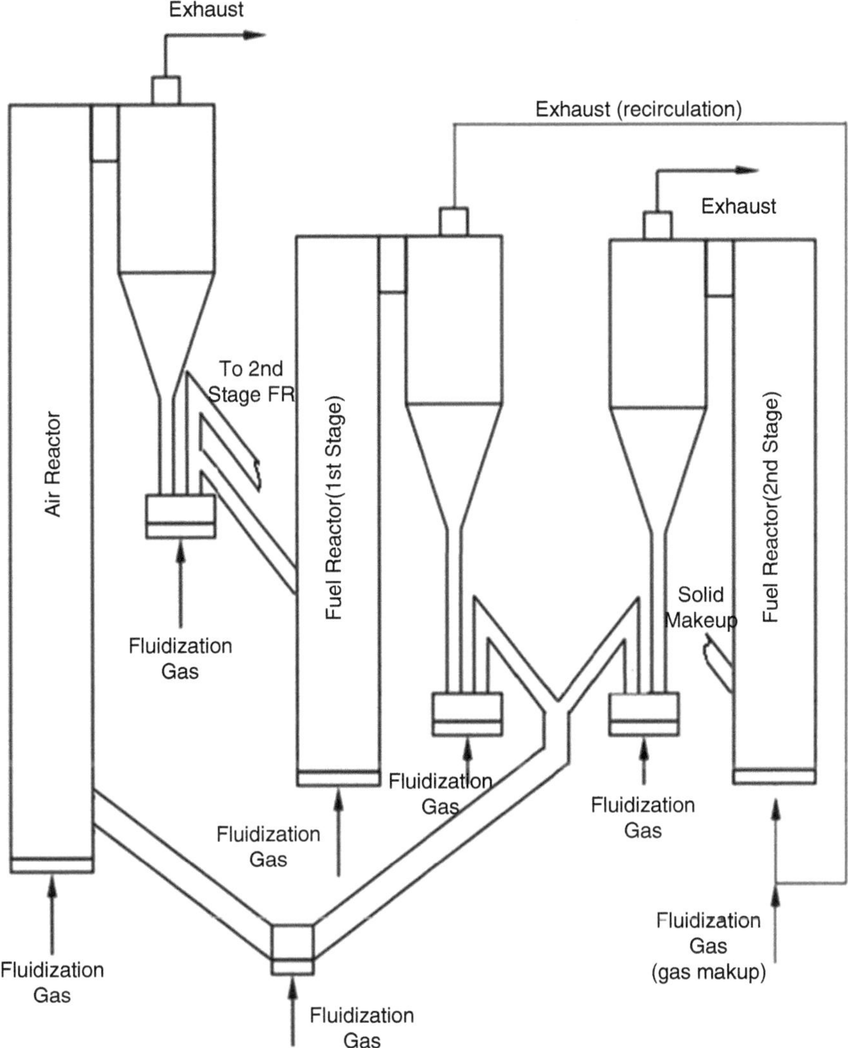

Fig. 3.9 Schematic of the proposed two-stage fuel reactor system for CLC

incorporate the effect of the slow rate of the gasification reaction into the process simulation. The RPLUG reactor is a plug flow rate reactor that simulates a rate-controlled reaction based on the reactor dimension, temperature, and the reaction kinetics provided. The reactor dimension is an independent variable that is varied in different multi-staging scenarios described in this section. The temperature of the gasifier is set at 1400 °C for all cases described following the work of Zhang et al. [10]. The kinetics for the gasification reaction is given in Table 3.10. The pre-exponent factor of 0.002 and the activation energy of 175,846 J/kmol are

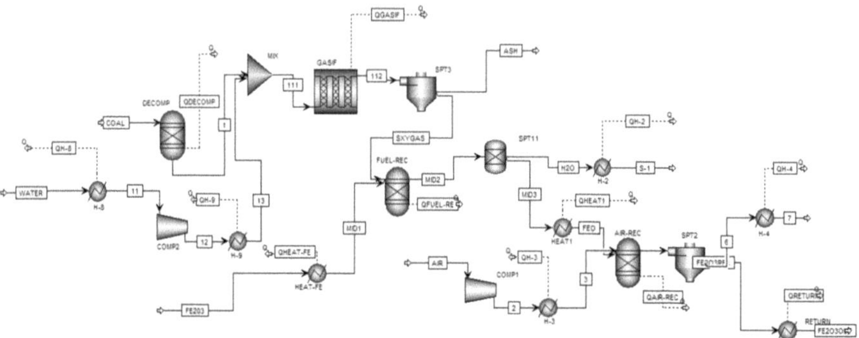

Fig. 3.10 Modified flow sheet of the single-stage CLC model in Aspen Plus

Table 3.10 Outline of the various reactors used in the single-stage CLC model in Aspen Plus

Name	Reactor type	Function	Chemical equation	Temperature (°C)	Pressure (atm)
DECOMP	RYIELD	Coal devolatilization	Coal → volatiles + C	–	–
GASIF	RPLUG	Char gasification	$C + H_2O \rightarrow CO + H_2$	1400	–
FUEL-REC	RGIBBS	Oxygen carrier reduction	$3Fe_2O_3 + CO/$ $H_2 \rightarrow 2Fe_3O_4 + CO_2/H_2O$	950	10
AIR-REC	RGIBBS	Oxygen carrier oxidation	$4Fe_3O_4 + O_2 \rightarrow 6Fe_2O_3$	935	10

used. The syngas produced in the gasification reaction is completely reacted with the Fe_2O_3 oxygen carrier in the fuel reactor to produce CO_2, H_2O, and Fe_3O_4. The fuel reactor in the model is the RGIBBS reactor, which considers all possible reactions and products and simulates chemical equilibrium by minimizing the Gibbs free energy. The fuel reactor is set at 950 °C and 10 atm. It should be noted that the RYIELD, RPLUG, and RGIBBS reactors together represent the fuel reactor in the physical setup. Aspen Plus cannot simulate the various reactions that take place in these three steps within a single reactor; therefore it is necessary to split the fuel reactor into three different Aspen reactor modules as described above. Therefore, the RGIBBS reactor represents the same physical volume as the RPLUG reactor, but since the oxygen carrier reduction reactions are significantly faster than the coal gasification reaction, the equilibrium model in the form of the RGIBBS reactor is adequate for the oxygen carrier reduction. After the fuel reactor, the Fe_3O_4 and any unreacted Fe_2O_3 are separated from the rest of the stream and transferred to the air reactor, which is also modeled as an RGIBBS reactor set at 935 °C and 10 atm. The rest of the stream is sent to a heat exchanger and released as exhaust. The coal used in the simulations conducted in this study is bituminous Colombian coal, which has physical and chemical properties as described in Table 3.1.

3.4.1 Single-Stage Fuel Reactor Model

For the Colombian coal considered in this section, the optimum ratio of coal, air, and Fe_2O_3 to maximize the energy output is 1:10:70 by mass. Aspen Plus simulations with the revised flow sheet shown in Fig. 3.10 are conducted using this optimum ratio for various dimensions of the RPLUG reactor block used to model the gasification reactor. The goal of these simulations is to isolate the effect of the relative gasifier volume parameter, which is representative of the relative fuel reactor volume, on the overall char conversion rate in the system. For each case, the net thermal energy output is recorded, and the chemical compositions of the fuel reactor and air reactor outlet streams are analyzed to determine the completeness of the respective reactions; the results are presented in Table 3.11 for various gasifier dimensions.

It should be noted that the dimensions of the gasifier in the Aspen Plus flow sheet in Fig. 3.10 simply provide a means of implementing the rate dependence of the char gasification on the relative performance of the CLC system as a function of the relative volume; they do not represent the absolute dimensions of an actual fuel reactor. Therefore, for simplicity, a length of 1 m and a diameter of 1 m are considered as the baseline reactor size, corresponding to a baseline volume of $\pi/4$ or 0.785 m^3. The term "relative volume" in Table 3.11 refers to a gasifier volume normalized by the baseline volume of 0.785 m^3. In test cases 5(a) and 5(b), the length and diameter of the gasifier are varied in turn while keeping the relative volume constant. The identical reaction completeness and net energy output in cases 5a and 5b show that the volume is the only parameter that affects the reaction rate in the gasifier. Therefore, for all other cases, the gasifier volume is varied by changing its length between 0.2 m and 2 m corresponding to a relative volume change between 20% and 200% of the baseline volume.

Table 3.11 Results of the single-stage fuel reaction model for CLC

Case #	1	2	3	4	5(a)	5(b)	6	7	8
Coal feed (kg/h)	100	100	100	100	100	100	100	100	100
Fe_2O_3 feed (kg/h)	3000	4000	5000	6500	7000	7000	7300	7300	7300
Gasifier diameter (m)	1	1	1	1	1	0.5	1	1	1
Gasifier length (m)	0.2	0.4	0.6	0.8	1	4	1.1	1.5	2
Gasifier relative volume	0.2	0.4	0.6	0.8	1	1	1.1	1.5	2
Char fraction reacted in gasifier	0.183	0.365	0.548	0.730	0.913	0.913	1.000	1.000	1.000
Fe_2O_3 residue in fuel reactor (kg/s)	0.197	0.175	0.154	0.271	0.111	0.111	0.051	0.051	0.051
Fe_3O_4 produced in fuel reactor (kg/s)	0.615	0.905	1.194	1.483	1.772	1.772	1.911	1.911	1.911
Net energy output (kW)	218.7	341.2	463.7	586.2	708.8	708.8	767.3	767.3	767.3

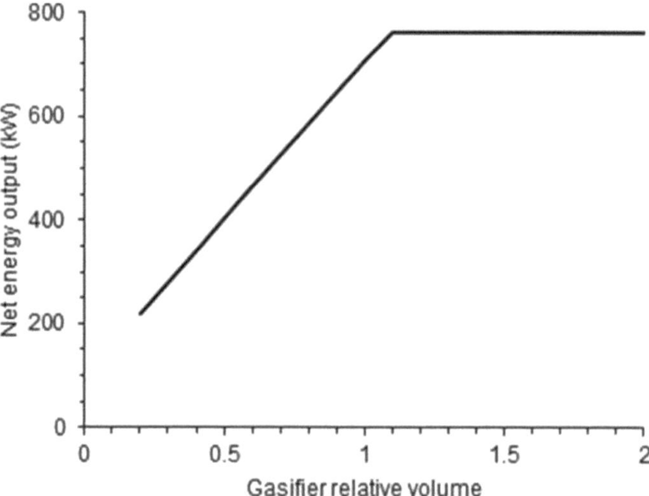

Fig. 3.11 Effect of gasifier volume on energy output in a single-stage fuel reactor model for CLC

In test cases 1–5, the gasification reaction is incomplete, as evidenced by the leftover carbon in the gasifier outlet stream. For these cases, the quantity of oxygen carrier in the system is reduced accordingly to avoid heating the excess oxygen carrier that is not used in extracting energy from the coal. On the other hand, in cases 6–8, since all the coal is consumed, the mass of air and Fe_2O_3 in the system are increased slightly to a ratio of coal vs. air vs. Fe_2O_3 by mass of 1:10.5:73 to ensure a small but not negligible amount of oxygen and Fe_2O_3 left in the outlet stream, thus eliminating the possibility of air or Fe_2O_3 being the limiting reactant in the CLC process. The simulation results of the single-stage fuel reaction model for different gasifier relative volumes are given in Table 3.11; the variation of net energy output with gasifier volume is plotted in Fig. 3.11.

From Fig. 3.11, it can be seen that the energy output increases linearly with an increase in the gasifier relative volume until the energy output reaches a maximum when the relative volume reaches 1.1, corresponding to an actual volume of 0.864 m^3. Any further increase in volume has no effect on the energy output, since all the char has already been gasified and the maximum energy in the coal has been released. This is a limitation of the Aspen Plus software as it does not take into account the additional heat required to bring the larger gasifier to the operating temperature. According to case 6 of Table 3.11, the gasifier with relative volume of 1.1 is the smallest reactor that is still big enough to completely react all the carbon in the coal, thus representing the optimal size of the gasifier.

3.4.2   Two-Stage Fuel Reactor Model

The flow sheet for the two-stage CLC model is shown in Fig. 3.12; it is an extension of the single-stage model. The difference in the two-stage CLC model lies in that a second fuel reactor consisting of an RPLUG reactor and an RGIBBS reactor is added after the first stage so that any unburned char from the first stage can be gasified and combusted in the second stage. It should be noted that the coal decomposition reactor and the cyclonic separator are only present in the first stage. The decomposition reactor is only needed in the first stage because the coal devolatilizes completely in a single stage. Similarly, the cyclonic separator is only used in the first stage because the coal ash only needs to be removed once after the coal decomposition.

The simulation results of the two-stage model for different gasifier relative volumes are presented in Table 3.12. It should be noted that the relative volume of the gasifier in each fuel reactor stage is kept the same. The char is partially gasified in the first gasifier and reacts with the Fe_2O_3 fed into the first fuel reactor. Any unreacted char is then transferred to the second gasifier where it is gasified either completely or partially depending on the gasifier size and reacts with the remaining Fe_2O_3. The ratio of the mass of Fe_2O_3 distributed to each first stage varies depending on the fraction of carbon being reacted in the first stage to avoid wastage. Since the gasifier in each stage has the same relative volume, the same fraction of char is gasified in each stage. If the first stage already gasifies over 50% of the char, the second stage gasifies the remainder. Accordingly, if less than 50% of the char is gasified in the first stage, the Fe_2O_3 is distributed equally to both stages, as in cases 9–13 in Table 3.12. Otherwise, the distribution of Fe_2O_3 across the two stages is in the same ratio as the char gasification.

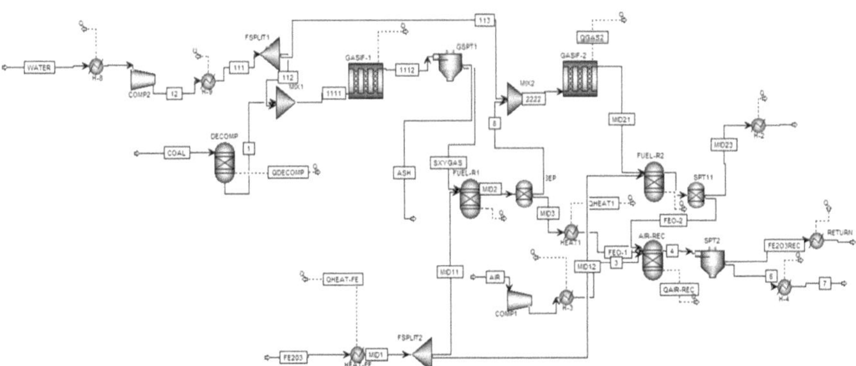

Fig. 3.12 Flow sheet of the two-stage fuel reaction model for CLC in Aspen Plus

Table 3.12 Results of the two-stage fuel reaction model for CLC

Case #	9	10	11	12	13	14	15	16	17	18
Gasifier relative volume	0.2	0.4	0.45	0.5	0.55	0.6	0.8	1	1.5	2
Total Fe_2O_3 feed (kg/h)	3500	6000	7300	7300	7300	7300	7300	7300	7300	7300
Fe_2O_3 fraction in stage 1	0.5	0.5	0.5	0.5	0.5	0.548	0.730	0.913	1.000	1.000
Net char fraction reacted after stage 1	0.183	0.365	0.411	0.456	0.502	0.548	0.730	0.913	1.000	1.000
Net char fraction reacted after stage 2	0.365	0.730	0.821	0.913	1.000	1.000	1.000	1.000	1.000	1.000
Net energy output (kW)	334.3	579.3	640.5	701.8	760.3	760.3	760.4	761.0	762.3	762.2

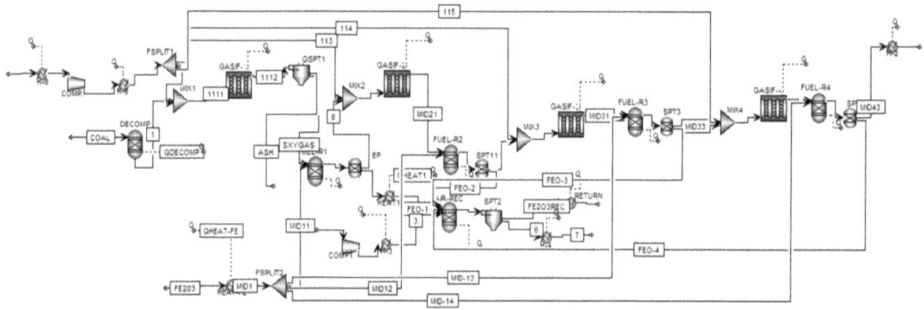

Fig. 3.13 Flow sheet of the four-stage fuel reaction model for CLC in Aspen Plus

3.4.3 Four-Stage Fuel Reactor Model

The flow sheet for the two-stage CLC process is extended to four stages by adding two more fuel reactor stages, as shown in Fig. 3.13. It is expected that the four-stage CLC simulation will confirm the observations from the two-stage model. The additional stages are identical to the second stage. The oxygen carrier is distributed evenly among the four stages, since the amount reacted in the later stages is always less than or equal to the amount reacted in the stage preceding it. The results of the four-stage model with varying relative volume of the gasifier are given in Table 3.13.

Figure 3.14 plots the net energy output against the gasifier volume in each stage for the two-stage and four-stage fuel reaction models and compares it with the single-stage setup. The gasifier volume follows a linear relationship with the net energy output in the two-stage and four-stage models until it reaches an optimum

Table 3.13 Results of the four-stage fuel reaction model for CLC

Case #	19	20	21	22	23	24	25
Gasifier relative volume	0.1	0.15	0.2	0.25	0.275	0.3	0.4
Total Fe_2O_3 feed (kg/h)	7300	7300	7300	7300	7300	7300	7300
Net char fraction reacted after stage 1	0.091	0.137	0.183	0.228	0.251	0.274	0.365
Net char fraction reacted after stage 2	0.183	0.274	0.365	0.456	0.502	0.548	0.730
Net char fraction reacted after stage 3	0.274	0.411	0.548	0.685	0.753	0.821	1.000
Net char fraction reacted after stage 4	0.365	0.548	0.730	0.913	1.000	1.000	1.000
Net energy output (kW)	334.3	456.8	579.3	701.7	760.6	760.3	760.3

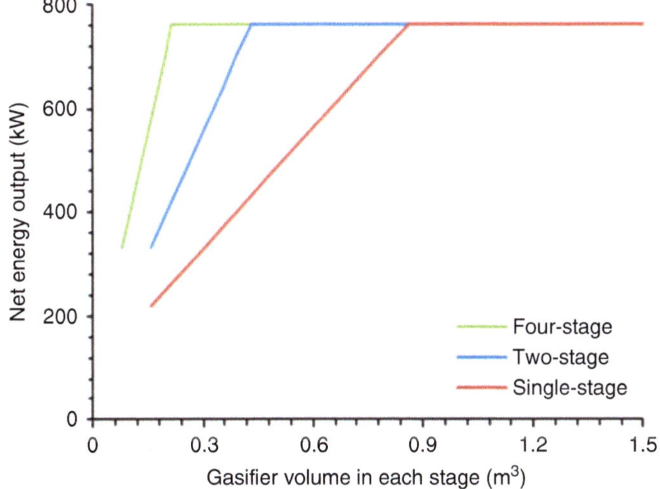

Fig. 3.14 Effect of gasifier volume on energy output in the two- and four-stage fuel reaction models for CLC

value. Increasing the gasifier volume further has negligible effect on the energy output. It is clear that the optimal relative volume for each gasifier in the two-stage model is 0.55, corresponding to an actual volume of 0.432 m^3, which yields a net energy output of 760.3 kW. The optimal gasifier relative volume for each gasifier in the two-stage model is exactly half of that in the single-stage model. Thus, the total volume of the fuel reactor across the two stages remains the same. The slight decrease in net energy output of 7 kW compared to the single-stage setup can be ascribed to rounding errors in the data and is not indicative of any losses due to multi-staging. The trends observed from the two-stage results are confirmed by the four-stage model—the optimum relative gasifier volume is 0.275, corresponding to an actual volume of 0.216 m^3, exactly one-fourth of the optimal volume of the single-stage case. The net energy output remains unchanged at 760.6 kW.

The results indicate that although the total reactor volume required to maintain complete char conversion remains the same, multi-staging allows a significant reduction in the volume of each reactor without sacrificing energy output. In other words, instead of having one large fuel reactor, multiple smaller fuel reactors can be used to burn the same amount of coal. The use of smaller reactors offers notable improvements in the fluidization behavior of the CLC fluidized beds that can be ascertained from hydrodynamic simulations using computational fluid dynamics software. Another interpretation is that multi-staging can offset the high coal residual time required in a single reactor of the same size by ensuring complete char conversion over multiple reactor stages. However, the energy required to heat multiple smaller fuel reactors to the operating temperature will be greater since the multiple reactors having the same cumulative volume will have a higher total surface area for heat loss. The cost of manufacturing and transporting multiple reactors versus a single large one should also be factored in this decision. Overall, this study demonstrates that multi-staged fuel reaction for CLC should be considered a viable alternative to single-stage CLC and warrants experimental investigation in the future.

3.4.4 Techno-Economic Analysis

To perform the techno-economic analysis of the simulations, the Aspen Plus Economic Analyzer function in Aspen Plus is employed. One of the needed inputs is to determine the cost of raw materials. In order to find the capital cost of the component type for each of the blocks in Fig. 3.9, there is a need to determine the type of equipment required for the components. The costs of single-stage, two-stage, and four-stage systems are considered. The techno-economic analysis of the three Aspen Plus models—single-stage, two-stage, and four-stage—has shown that CLC is an economically viable system to use as a replacement of the boiler in a traditional coal power plant.

The results also show that the capital costs of the system increase with increase in number of stages as shown in Fig. 3.15. The difference in the cost between the two-stage and single-stage model is ~ $130,000, and the difference in cost between the two-stage and four-stage model is ~ $780,000. This difference can be attributed to two reasons. The first reason is that even though the total volumes of equipment in the models of all the stages are the same, two stages and four stages still require more equipment (although of different size) than the single stage. Although the equipment gets smaller in two and four stages, the cost of the material required in the equipment for additional stages still creates the difference in price. The second reason for the difference is the utility infrastructure. Since the amount of equipment increases for the two-stage and four-stage CLC models, they require excess wiring for electricity and extra piping which results in an increase in the capital cost.

The results for the operating costs of the system are almost opposite to that for the capital cost as shown in Fig. 3.16. The increase in number of reactors brings down the operating cost of the systems. The cost of the two-stage model is ~ $210,000 less than the single-stage model, and the cost of the four-stage model is ~ $100,000 less than that of the two-stage model. This difference in economic cost directly correlates

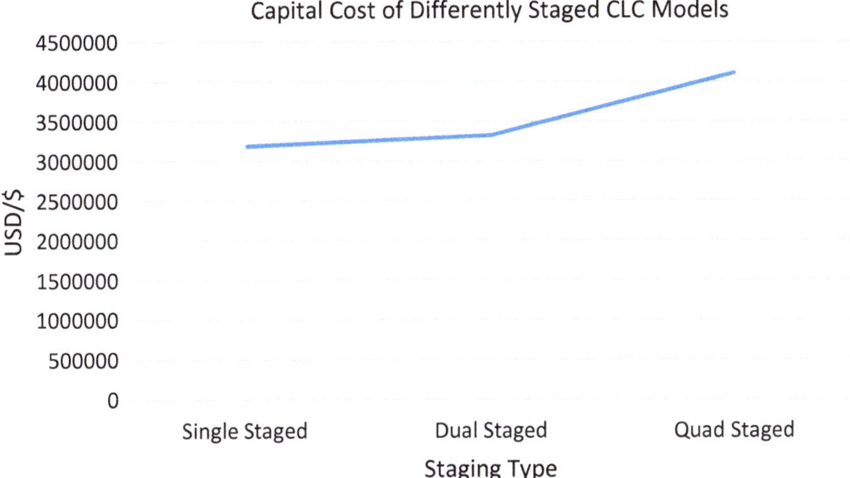

Fig. 3.15 Capital cost of the three differently staged CLC models

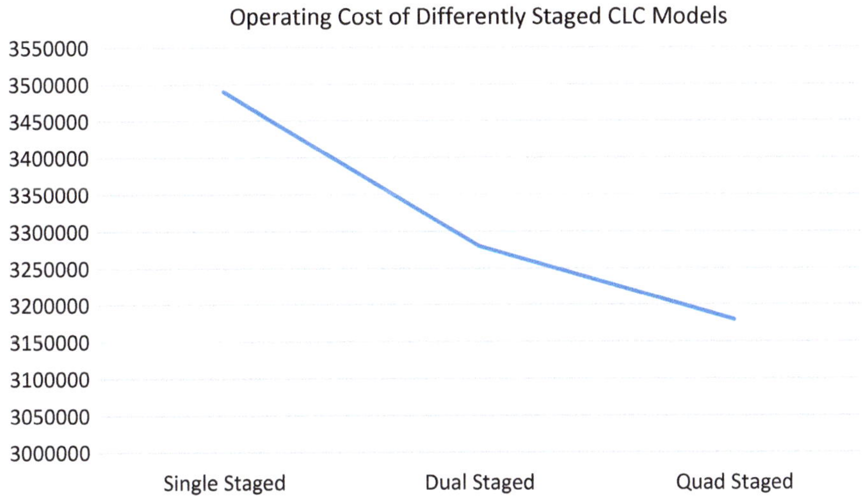

Fig. 3.16 Graph of the operating cost of differently staged CLC models

with the efficiency that multi-staging provides. Multi-staging is able to produce energy more efficiently from the coal, and thus it requires less raw material to produce the same amount of energy as a single-stage system.

When evaluating the economic viability of multi-staged CLC systems, the capital cost and the operating cost are the two significant factors. The two-stage system is more economically viable than a single-stage system, although the two-stage system has a higher capital cost of ~ $130,000; however the two-stage system costs ~ $210,000 less in operating cost. The four-stage system on the other hand is not an economically viable system since not only it costs $780,000 more than the two-stage system in capital costs; the difference in operating costs between the two-stage and four-stage systems is comparatively marginal compared to the four-stage system costing only $100,000 less than the two-stage system.

3.5   iG-CLC Process Simulation with a Mixture of Biomass and Coal

This section aims at further contributing to the knowledge of CLC by using biomass and a mixture of biomass and coal as a fuel and conducting process simulation using Aspen Plus. First, the ASPEN Plus process simulation results are validated by using the experimental data from a 0.5 kWth CLC reactor, which employed three types of biomass, namely, pine sawdust, almond shells, and olive stones. Then the effects of using Mn_2O_3 and Fe_2O_3 as oxygen carrier are simulated and compared. The influence of the fuel reactor temperature on the CLC performance is evaluated which includes gas concentrations, carbon capture efficiency, and power output. Finally, a mixture of biomass and coal is considered with different percentages by mass, and its performance is evaluated in terms of gas concentrations, carbon capture efficiency, and power output.

3.5.1   Process Simulations of Pure Biomass

The experimental setup employed by Mendiara et al. [11] shown in Fig. 3.17 has been employed to establish a corresponding process simulation model and flow sheet in Aspen Plus. The experimental setup consists of two fluidized bed reactors connected by a loop seal to prevent the mixing of gases between the two fluidized bed reactors. It should be noted that at the beginning of the experiment, air and N_2 were used as fluidizing agents during the heating period. After reaching the desired temperatures, the fluidizing agent in the fuel reactor was switched to steam, and then the biomass was fed into the reactor. Three biomass types—pine sawdust, almond shells, and olive stones—were used in the experiment.

In Aspen Plus, the biomass was modeled as a nonconventional solid using the ultimate and proximate analyses provided in Mendiara et al.'s experiment [11]. These values are shown below in Table 3.14.

In Mendiara et al.'s experimental setup [11], an iron ore known as Tierga was used as the oxygen carrier. Based on data provided in Mendiara et al.'s experiment, Tierga ore is primarily hematite (Fe_2O_3), consisting of 76.5% hematite by mass with the remainder of the ore being a mixture of silica (SiO_2), aluminum oxide (Al_2O_3), calcium oxide (CaO), and magnesium oxide (MgO) [12]. The oxygen carrier (OC) stream in the simulation included these impurities in order to simulate the

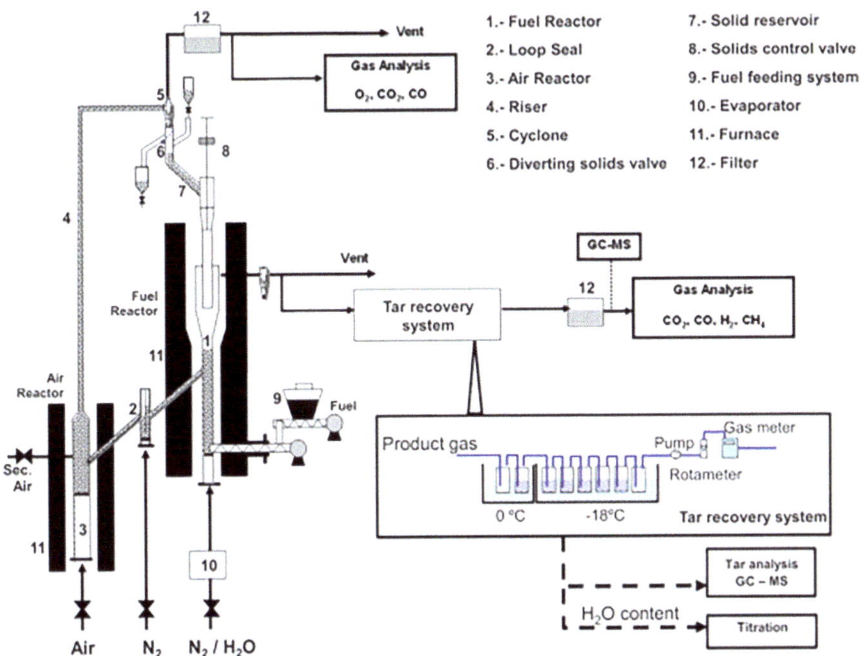

Fig. 3.17 Schematic of the apparatus used in Mendiara et al.'s experiment [11]

Table 3.14 Ultimate and proximate analysis of the three biomass types [11]

	Pine sawdust	Olive stone	Almond shell
Ultimate analysis (wt. %)			
Carbon	51.3	46.5	50.2
Hydrogen	6.0	4.8	5.7
Nitrogen	0.3	0.2	0.2
Sulfur	0.0	0.0	0.0
Oxygen	37.8	38.3	40.5
LHV (kJ/kg)	19,158	16,807	18,071
Proximate analysis (wt. %)			
Moisture	4.2	9.4	2.3
Ash	0.4	0.8	1.1
Volatile matter	81.0	72.5	76.6
Fixed carbon	14.4	17.3	20.0

experiment as accurately as possible. In addition, the biomass and OC flow rates were varied in Mendiara et al.'s experiment [11] to maintain optimal OC to fuel ratio. Table 3.15 shows the complete input values for each fuel reactor temperature and biomass type. For all simulations described in this section, the air reactor temperature was 950 °C.

Table 3.15 Input values in Aspen Plus corresponding to the values used by Mendiara et al. [11]

Parameters	Pine sawdust				Olive stone			Almond shell		
Temperature (°C)	895	910	955	985	905	955	980	905	955	985
Biomass (g/h)	152	113	98	98	164	141	141	190	102	88
OC (kg/h)	10.1	9.2	6.7	6.7	8.8	7.6	7.6	13	7	5.9

Table 3.16 Process models used in various parts of the CLC process in Aspen Plus

Name	Model	Function	Reaction
DECOMP-1	RYIELD	Biomass devolatilization and pyrolysis	Biomass $\rightarrow$ volatile matter + char
RGIBBS	RGIBBS	Gasification	Char + volatile matter $\rightarrow$ $CO_2 + H_2O$
FUEL-R	RSTOIC	OC reduction reactions and further combustion [11]	Char (mainly C) + $H_2O \rightarrow$ $H_2 + CO$ Char (mainly C) + $CO_2 \rightarrow$ 2 CO $CO + Fe_2O_3 \rightarrow CO_2 + 2\ Fe_3O_4$ $H_2 + 3\ Fe_2O_3 \rightarrow H_2O + 2\ Fe_3O_4$ $CH_4 + 12\ Fe_2O_3 \rightarrow$ $2\ H_2O + CO_2 + 8\ Fe_3O_4$ $CO + H_2O \leftrightarrow CO_2 + H_2$
AIR-R	RSTOIC	OC oxidation reaction [13]	$0.5\ O_2 + 2\ Fe_3O_4 \rightarrow 3\ Fe_2O_3$ C (unconverted char) + $O_2 \rightarrow$ CO_2

In Aspen Plus, the flow sheet is set up as "solids with metric units," and the MIXCINC stream class is assigned to the simulation. The MIXCINC stream class allows for both conventional and nonconventional streams to be used in the simulation. The nonconventional stream class is used to model the biomass, while the conventional stream class is used to model all other streams in the simulation. For simplicity, the IDEAL base calculation method is selected; this allows for calculations to be done using the ideal gas law. Due to inherent limitations of the Aspen Plus software, multiple blocks are used to model the fuel reactor. For the fuel reactor, a RYIELD reactor is first used to model the pyrolysis and devolatilization of the fuel. The products of the RYIELD reactor are then moved to RGIBBS reactor which is used to model the gasification of the decomposed fuel. The products of the RGIBBS reactor are then moved to RSTOIC reactor which is used to simulate the fuel combustion. These three blocks together represent the fuel reactor. The air reactor is modeled as a single RSTOIC reactor. An overview of these components and their respective functions is shown in Table 3.16. The model constructed in Aspen Plus is shown in Fig. 3.18.

After the validation of process simulations against the experiment of Mendiara et al. [11], the simulations were conducted for the evaluation of the performance of CLC setup by using different oxygen carriers. For this purpose, manganese (III) oxide was used as the oxygen carrier instead of hematite. Mn-ores also possess a

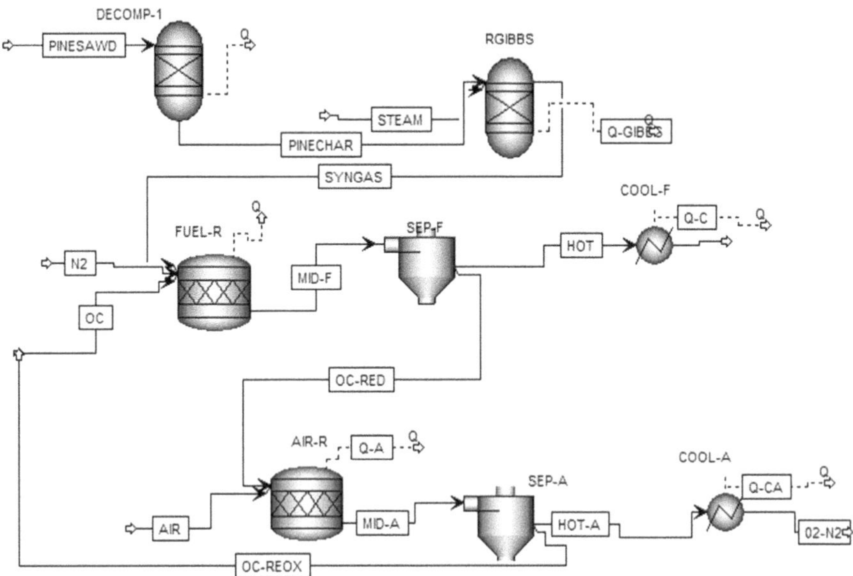

Fig. 3.18 Flow sheet of experimental setup in Aspen Plus

relatively high reactivity, which makes them a potential alternative to Fe-based oxygen carriers. In Aspen Plus, the composition of the oxygen carrier stream was changed to be primarily Mn_2O_3 but still included the impurities that were present in the Fe_2O_3 oxygen carrier stream. The reactions in the fuel and air reactors were changed to use the Mn-based OC in place of the Fe-based OC. The stoichiometry of these reactions remained unchanged. By changing the OC, the concentrations of gases in the fuel reactor were compared to the simulations that had employed Fe_2O_3 as the oxygen carrier. In addition, the power output of the two systems at various temperatures was compared, As previously mentioned, the main focus of the process simulations for biomass is its validation against the experimental data obtained by Mendiara et al. [11] In their experiment, biomass was combusted in the fuel reactor at various temperatures, in order to determine the optimal temperature to capture the purest stream of CO_2.

As summarized in Table 3.17, the concentrations of various compounds that were determined by Mendiara et al. [11] are compared with the concentrations obtained from the simulations for three kinds of biomass. Methane is the compound that shows the greatest discrepancy between the experimental data and the simulation results, and its concentration continues to decrease as the fuel reactor temperature increases. However, the concentrations of other products rise above the experimentally observed value. The concentrations of these products do not deviate from the experimental values by a substantial amount, with the concentrations of CO_2 being within 2% of the experimentally observed concentration. This indicates that the results of the simulation are reasonably accurate validation of the experimental results [11].

Table 3.17 Validation of simulation results against experimental results

Pine sawdust

Temperatures	895 °C		910 °C		950 °C		985 °C		Average error
Gas composition	vol %	Error	vol %	Error	vol %	Error	vol %	Error	
CO_2 simulation	69.7	0.7%	71.7	0.8%	72.6	0.7%	67.4	3.4%	1.4%
CO_2 experiment	69.2		71.1		72.1		65.2		
CO simulation	11.2	1.8%	9.6	1.1%	10.7	5.9%	13.1	12.0%	5.2%
CO experiment	11.0		9.5		10.1		11.7		
H_2 simulation	11.7	5.4%	12.5	1.6%	12.5	9.6%	16.7	8.4%	6.3%
H_2 experiment	11.1		12.3		11.4		15.4		
CH_4 simulation	7.3	−16.1%	6.2	−12.7%	3.9	−39.1%	2.7	−64.9%	−33.2%
CH_4 experiment	8.7		7.1		6.4		7.7		

Olive stone

Gas composition	895 °C		905 °C		955 °C		980 °C		Average error
	vol %	Error	vol %	Error	vol %	Error	vol %	Error	
CO_2 simulation			71.3	1.9%	68.0	2.7%	67.0	3.1	2.6%
CO_2 experiment			70.0		66.2		65.0		
CO simulation			11.2	8.7%	14.1	11.0%	14.2	13.6	11.1%
CO experiment			10.3		12.7		12.5		
H_2 simulation			12.2	8.0%	14.9	13.7%	16.4	13.1	11.6%
H_2 experiment			11.3		13.1		14.5		
CH_4 simulation			4.9	−41.7%	2.9	−63.8%	2.2	−72.5%	−59.3%
CH_4 experiment			8.4		8.0		8.0		

Almond shell

Gas composition	895 °C		905 °C		955 °C		980 °C		Average error
	vol %	Error	vol %	Error	vol %	Error	vol %	Error	
CO_2 simulation			71.7	1.3%	66.7	2.6%	74.1	1.6%	1.8%
CO_2 experiment			70.8		65.0		72.9		
CO simulation			11.4	9.6%	12.1	14.2%	11.1	13.3%	12.3%
CO experiment			10.4		10.6		9.8		
H_2 simulation			11.1	6.7%	18.0	11.1%	12.3	8.8%	8.9%
H_2 experiment			10.4		16.2		11.3		
CH_4 simulation			5.8	−31.0%	3.2	−61.0%	2.5	−58.3%	−50.1%
CH_4 experiment			8.4		8.2		6.0		

In the experiment, the raw biomass materials were dried and sieved to desired size to obtain a particle distribution between 0.5 and 2 mm and then mixed with the oxygen carrier [11]. Due to good mixing between oxygen carrier and small biomass particles, some residual char gasification took place in the upper part of the fuel reactor [13]. However, there were not sufficient fresh oxygen carrier particles to react with the gaseous products from gasification of residual char near the bed surface of the fuel reactor. Inevitably, some CO and CH_4 would therefore exist in the flue gas of the fuel reactor, especially CH_4. This is the same phenomenon that was observed both in a 10 kWth and a 0.5 kWth CLC unit [13]. However, in the process simulation, the CH_4, CO, and H_2 will have reaction with carrier particles only if there are enough carrier particles; thus the predicted CH_4 concentration in simulation is lower than that in the experimental resulting in large discrepancy between the simulation and experimental results for CH_4. Therefore, it can be concluded that the simulation model for methane production in Aspen Plus needs further improvement to give a better prediction.

1. Gas concentrations from pine sawdust

For pine sawdust, the biomass was fed into the fuel reactor at 895, 910, 950, and 985 °C, while the concentrations of various gases, (viz., CO_2, CO, H_2, and CH_4) were measured in the stream leaving the fuel reactor. The simulations most closely model the experimental data at lower temperatures, while deviating at higher temperatures. CO_2 concentration is found to increase from 895 °C until 950 °C at which point the CO_2 concentration reaches its highest value and then begins to decrease. Overall, the concentrations of CO_2, CO, and H_2 in the fuel reactor agree fairly well with the experimental results for pine sawdust, as shown in Fig. 3.19. Biomass char gasification gets enhanced at higher fuel reactor temperatures thereby significantly reducing the char particles and CO_2 in the fuel reactor. As such, the purest CO_2 stream is obtained at a fuel reactor temperature of 950 °C. This is consistent in both the experimental results as well as in the simulations.

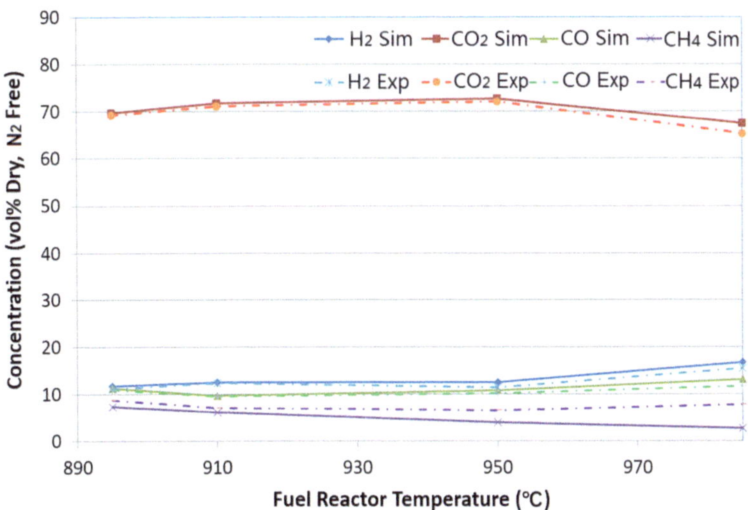

Fig. 3.19 Gas concentrations in fuel reactor from pine sawdust compared to experimental results obtained by Mendiara et al. [11]

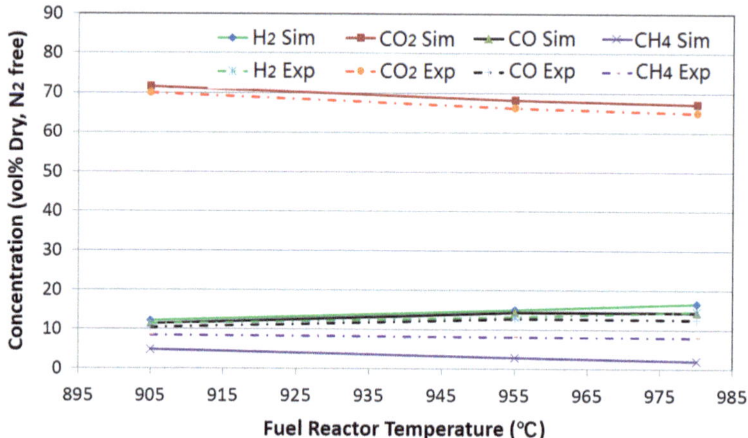

Fig. 3.20 Gas concentrations in fuel reactor from olive stones compared to experimental results obtained by Mendiara et al. [11]

2. Gas concentrations from olive stone

For olive stone, the biomass was fed into the fuel reactor at 905, 955, and 980 °C. The concentrations of H_2, CO, CO_2, and CH_4 are shown in Fig. 3.20. As was the case with the pine sawdust, methane is still underproduced by the simulation. The concentration of other gases also did not match as well as they did in the case of pine sawdust; however, the largest deviation was still only 2.0% (CO_2 at 980 °C), which indicates that the model constructed in Aspen Plus was an accurate representation of the experimental setup. It is shown in Fig. 3.20 that CO_2 concentration is the highest at 905 °C, and the concentration decreases gradually with the increase in temperature. It is also observed that the presence of high temperature improves the biomass gasification and generates a higher fraction of CO and H_2.

3. Gas concentrations from almond shell

For almond shell, the biomass was fed into the fuel reactor at 905, 955, and 985 °C. The concentrations of H_2, CO, CO_2, and CH_4 are shown in Fig. 3.21. Once again, the simulated value of concentration of methane is lower than the experimental value. The simulation model of methane production in Aspen Plus needs further improvement to give a better prediction. However, the data obtained from the simulation model fits the experimental data fairly well, with the largest deviation in concentration being 1.8% (H_2 at 955 °C). The CO_2 concentration initially decreases with temperature, until it reaches its lowest value at 955 °C. From this point onward, the CO_2 concentration increases as the temperature approaches 985 °C as shown in Fig. 3.21.

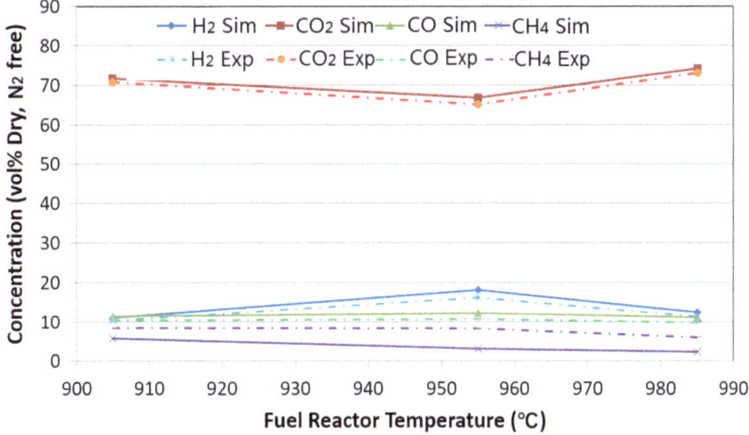

Fig. 3.21 Gas concentrations in fuel reactor from almond shell compared to experimental results obtained by Mendiara et al. [11]

4. Carbon capture efficiency (N_{cc})

One method of evaluating the performance of a CLC system is to determine its carbon capture efficiency (N_{cc}). The equation to determine the carbon capture efficiency is given below [11]:

$$N_{cc} = \frac{[F_{CO2,FR} + F_{CO,FR} + F_{CH,FR}]_{out}}{[F_{CO2,F\,R} + F_{CO,FR} + F_{CH4,FR} + F_{CO2,AR}]_{out}} \tag{3.3}$$

where $F_{x,FR}$ is used to denote the volume percentage of gas x in the fuel reactor, while $F_{CO2,\,AR}$ is used to denote the volume percentage of CO_2 in the air reactor. CO_2 in the air reactor is a byproduct of residual char combusted in the air reactor. Since a CLC system consists of interconnected fluidized beds, some char is carried from the fuel reactor into the air reactor by OC particles. At higher temperatures, more char is converted into carbonaceous gas in the fuel reactor, and as such, increasing the temperature of the fuel reactor decreases the amount of char that can reach the air reactor. Since gasification process is the rate-limiting process in CLC, higher fuel reactor temperature generally tends to accelerate the gasification of biomass. With the increase of temperature, the generation of volatile matter and syngas increases. Thus, increasing the fuel reactor temperature decreases the concentration of CO_2 in the air reactor and thereby increases the carbon capture efficiency. For both pine sawdust and almond shell, the carbon capture efficiency reaches approximately 100% at temperatures equal to or greater than 950 °C, while olive stones reach N_{cc} value of approximately 100% at 980 °C. This difference could be due to the fact that pine sawdust and almond shell have greater volatile matter content than olive stones, (81.0% and 76.6% compared to 72.5%, respectively). The results of computed carbon capture efficiency as a function of the fuel reactor

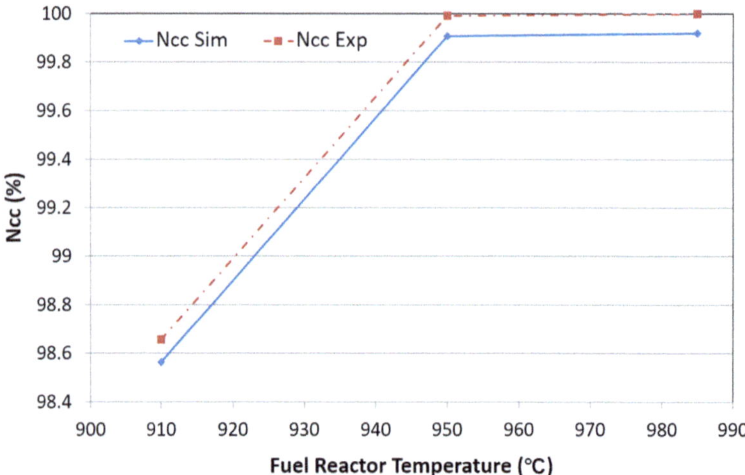

Fig. 3.22 Carbon capture efficiency for pine sawdust

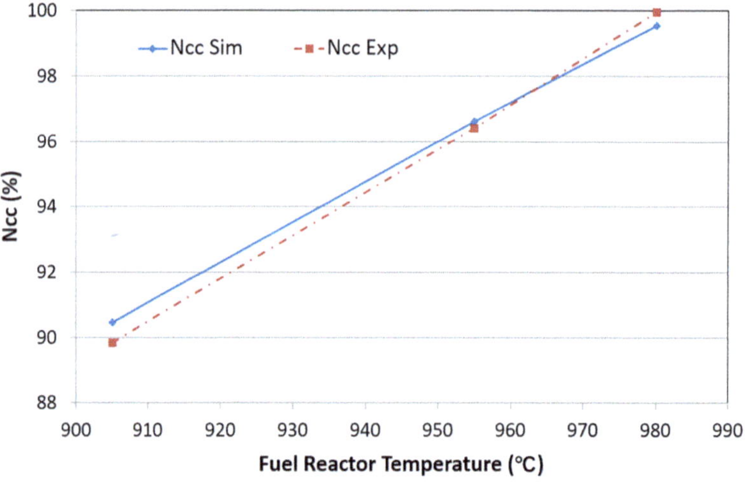

Fig. 3.23 Carbon capture efficiency for olive stones

temperature match the experimental results obtained by Mendiara et al. [11] quite closely as shown in Figs. 3.22, 3.23, and 3.24. Pine sawdust is the biomass type for which the simulation data fitted the experimental data most closely since the small particles of pine sawdust burn more thoroughly than olive stones and almond shell.

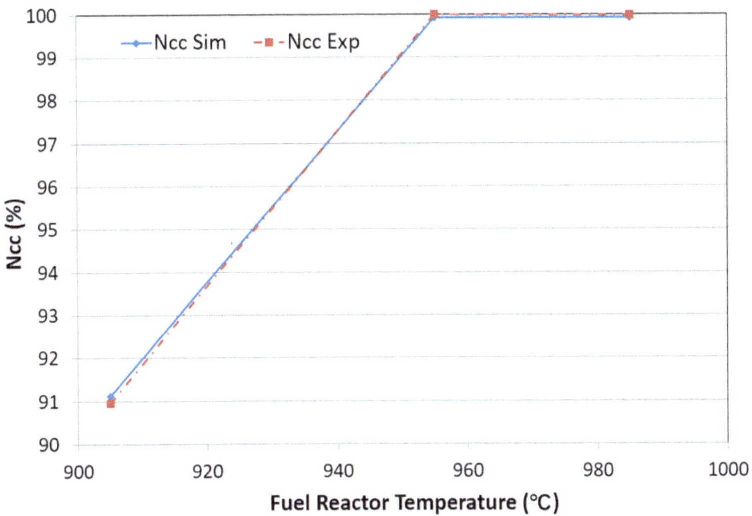

Fig. 3.24 Carbon capture efficiency for almond shell

3.5.2   *Effect of Different Oxygen Carriers*

After validating the process simulation model against the experimental data obtained by Mendiara et al. [11], the OC was changed from Fe_2O_3 to Mn_2O_3. Pine sawdust was used as the biomass of choice because the gas concentrations in the simulations for this biomass most closely matches the experimental results.

In changing the OC from a Fe-based compound to a Mn-based compound, the concentrations of CO and H_2 in the fuel reactor decreased slightly, while the concentration of CO_2 increased. The CH_4 value remained essentially unchanged. This indicates that Mn_2O_3 is more reactive as an OC and converts more CO to CO_2 than a hematite-based OC. However, since the maximum CO_2 concentration from the setup using Mn_2O_3 is only about 0.8% greater than what was obtained from Fe_2O_3, the use of a Mn-based OC in place of a Fe-based OC does not provide significant benefits. The carbon capture efficiency remains identical between the two oxygen carriers, but since these values are essentially 100% at temperatures greater than 950 °C, this is just an indication that Mn_2O_3 is equally effective at converting residual char to carbonaceous gases as Fe_2O_3. The concentrations of CO_2, CO, H_2, and CH_4 in the fuel reactor are shown in Figs. 3.25, 3.26, 3.27, and 3.28. The results of the simulation using Mn_2O_3 are compared directly with the results from the simulation using Fe_2O_3 instead of the experiment using that OC, in order to most clearly show the effect of changing the OC. It was found that switching the oxygen carrier from Fe_2O_3 to Mn_2O_3 caused the concentrations of CO and H_2 in the fuel reactor to decrease slightly, while the concentration of CO_2 increased slightly. Furthermore, changing the OC to Mn_2O_3 had no effect on the concentration of CH_4.

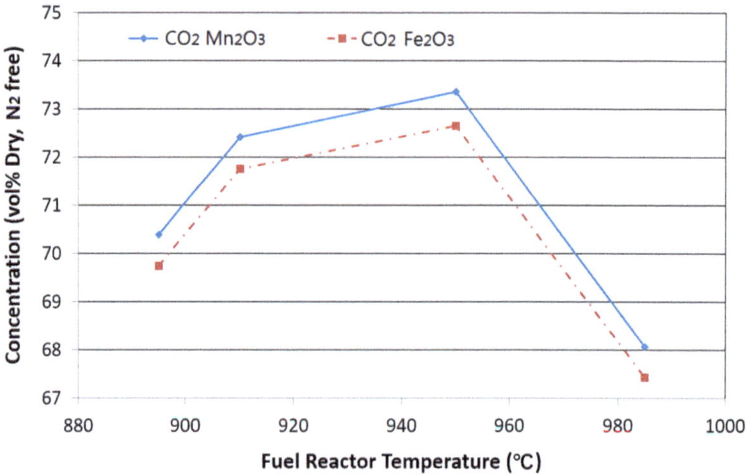

Fig. 3.25 CO_2 yield from Mn-based OC compared to Fe-based OC

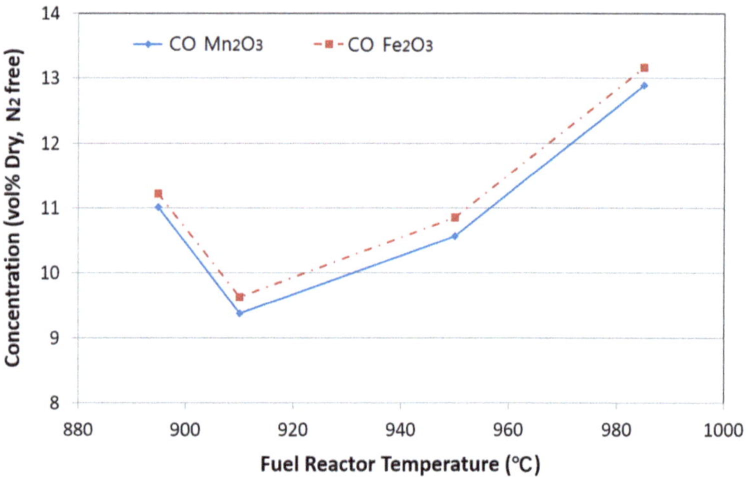

Fig. 3.26 CO yield from Mn-based OC compared to Fe-based OC

3.5.3   Power Output of Mixture of Biomass and Coal

1. Pure biomass

To further investigate both the utility of this CLC system and the effect of using Mn_2O_3 in place of Fe_2O_3, the power output of the system was determined at different temperatures, as can be seen in Fig. 3.29. Once again, the biomass of choice was pine sawdust. To maintain the consistency in the simulation model, biomass flow was

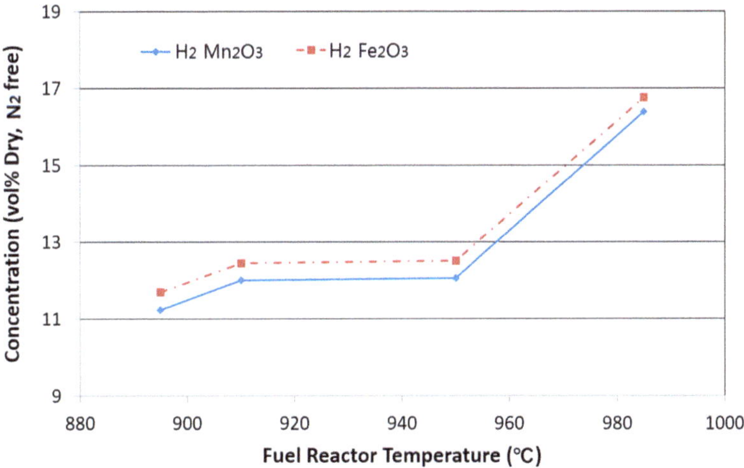

Fig. 3.27 H_2 yield from Mn-based OC compared to Fe-based OC

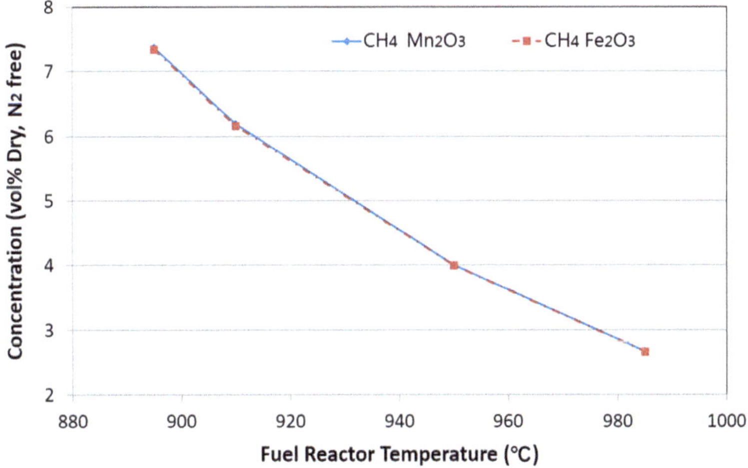

Fig. 3.28 CH_4 yield from Mn-based OC compared to Fe-based OC

decreased with temperature as specified in Mendiara et al.'s [11] paper. For both situations, power output falls sharply as temperature increases, and biomass flow rate decreases. Additionally, it can be seen that the CLC system using Fe_2O_3 displays a higher power output than the system using Mn_2O_3. At 895 °C, the power output is at its maximum value in both systems, at approximately 2220 W using Fe_2O_3 and approximately 1531w using Mn_2O_3. In addition, it can be observed that the power output falls to negative values for both scenarios once the temperature in the fuel reactor reaches 950 °C. This is an indication that the energy released from the

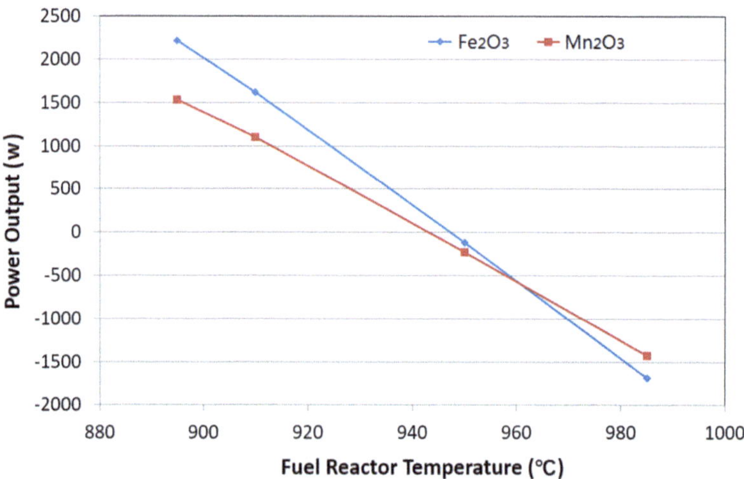

Fig. 3.29 Power output from pine sawdust using Fe-based OC and Mn-based OC

combustion of the biomass at that particular flow rate (98 g/h) is not sufficient to make up for the energy required to heat the fuel reactor to 950 °C and beyond.

2. Mixture of coal and biomass

A third case is presented to show how the power output varied between the two oxygen carriers when a mixture of coal and biomass is used in place of pure biomass. For this part of the simulations, bituminous coal is used. The ultimate and proximate analysis of the bituminous coal is shown below in Table 3.18 [14].

In addition to using pure biomass to compare the energy output when Fe_2O_3 is used as an OC to when Mn_2O_3 is used, the power output from a mixture of biomass and coal is also considered. The fuel reactor temperature is set at 895 °C, and an OC flow rate of 10.1 kg/h is used. The total biomass flow rate always added up to 152 g/h.

Figure 3.30 shows the power output achieved from a mixture of biomass and coal when Fe_2O_3 is used as the OC and when Mn_2O_3 is used as OC. It can be seen that the maximum power output achieved with Mn_2O_3 is substantially lower than the maximum power output achieved with Fe_2O_3 as OC. In addition, as the fraction of coal in the fuel stream increases, the power output also increases. This is due to coal's higher LHV, meaning that more heat is released when coal is combusted than when pure biomass is combusted. Since use of Mn_2O_3 as OC resulted in the system producing far less power compared to that with Fe_2O_3 as OC and since it did not lead to purer streams leaving the fuel reactor, it appears that Mn_2O_3 may not be a suitable alternative to Fe_2O_3 as an oxygen carrier.

Table 3.18 Ultimate and proximate analysis of bituminous coal [14]

Ultimate analysis (wt. %)	
Carbon	69.75
Hydrogen	4.3
Nitrogen	1.03
Sulfur	0.52
Oxygen	13.81
LHV (kJ/kg)	27,100
Proximate analysis (wt. %)	
Moisture	6.01
Ash	4.76
Volatile matter	35.1
Fixed carbon	54.13

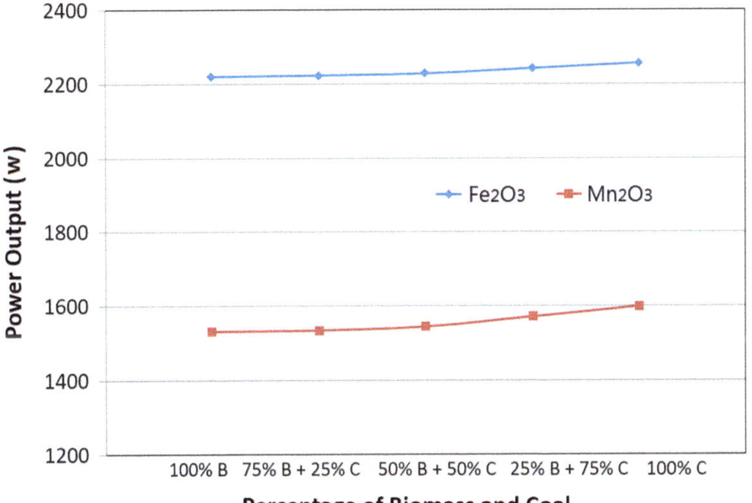

Fig. 3.30 Power output from mixture of biomass and coal using Fe-based OC and Mn-based OC

3.6   Calcium Looping Process

The calcium looping (CL) process consists of two interconnected reactors called the carbonator and the calciner in which the carbonation and calcination reactions take place respectively. The overall carbonation-calcination equilibrium reaction is given by

$$CaO\ (s) + CO_2\ (g) \rightleftharpoons CaCO_3(s) \tag{3.4}$$

The carbonation reaction entraps the CO_2 from the flue gas stream using the calcium oxide sorbent to form calcium carbonate ($CaCO_3$). The flue stream exiting

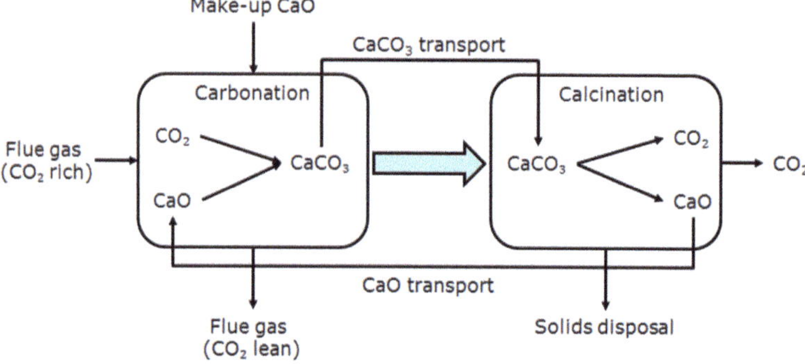

Fig. 3.31 Schematic representation of a calcium looping system with interconnected reactors

the carbonator is CO_2-lean and can be exhausted into the atmosphere in the case of post-combustion CaL or used for clean combustion in the case of pre-combustion CaL. The solid $CaCO_3$ from the carbonator is transported to the calciner where it is heated to decompose into CaO and CO_2.

The calciner produces a stream of pure CO_2 that is sent for pressurized storage for subsequent sequestration or use. The CaO is transported back into the carbonator to complete the loop. The typical setup for CaL mirrors that of a chemical looping combustion (CLC) plant, with the carbonator and calciner analogous to the air and fuel reactors in CLC and the $CaO/CaCO_3$ equivalent to the oxidized/reduced metal oxide oxygen carrier. A schematic representation of the CaL process is shown in Fig. 3.31. The make-up CaO flow and solids disposal are required to maintain reactivity of the sorbent; if the CaO is not replenished, the reaction rates would degenerate over time as the loop runs its course.

The energy penalty is often used to characterize the performance of a carbon capture system. In the context of CaL, the energy penalty is a measure of the portion of energy consumed by the carbonator and calciner and the transport of solids between the two reactors compared to the total energy released by the combustion process. The goal of a carbon capture process is to consume the least amount of energy while achieving a high CO_2 capture efficiency. Therefore, the estimation of the energy penalty in CaL is of great interest in the field of CCUS.

In this section, process models of calcium looping are developed using Aspen Plus to determine the energy penalty in a CaL system. The two main types of calcium looping systems are studied—post-combustion capture and pre-combustion capture—and their performances in terms of the energy penalty are compared. It should be noted that only thermal energy analysis is performed; the turbines and generators in the power plant are not included. Since heat energy does not transform into other forms of energy such as mechanical or electrical energy, the term "heat" is used interchangeably with "energy" in this paper. Aspen Plus also does not provide any means of calculating the energy required to transport the solids between the carbonator and the calciner and back. However, from the previous work on the solids transport in CLC, the energy penalty of the transport process is typically less than 1% of the energy released by combustion.

3.6.1 Calcium Looping with Post-combustion Capture

In post-combustion capture, the carbonator and calciner are included downstream of the combustion process to capture the CO_2 from the flue gases generated by the combustion of coal. In order to investigate the energy penalty associated with a calcium looping system for post-combustion capture, the overall heat production from a power plant without and with calcium looping must be determined. In the simulation, all inlet materials are set at room temperature, and the inlet coal properties are set as received rather than using those of dry coal.

3.6.1.1 Process simulation Setup

The materials used in the simulation include conventional and nonconventional components. Pure materials, which include all possible simple substances and chemical compounds comprising the elements C, N, O, H, S, and Cl that might be produced during the chemical reactions are designated as conventional materials. Properties for conventional materials are obtained from the Aspen Plus data bank. Mixtures such as coal and ash are designated as nonconventional solids.

1. Combustor

The work of Sivalingam [15] is used as a basis for the process models of calcium looping in this chapter. Illinois #6 coal is used in the current simulation to match the coal that was used in the simulation by Sivalingam [15]. Since coal is designated as a nonconventional solid in Aspen Plus, its attributes are defined based on its physical and chemical properties. The RYIELD reactor block in Aspen Plus is employed to decompose the nonconventional coal material into its constituent conventional materials to be able to simulate their reactions. The products of decomposition are the simple components—Ash, H_2O, C, H_2, N_2, Cl_2, S, and O_2. Mass percentages for the component yields are set based on the proximate and ultimate analysis of the Illinois #6 coal obtained from the work of Sivalingam [15] and summarized in Table 3.19.

The material stream from the RYIELD reactor goes into a burner, modeled as a RGIBBS reactor, along with air for combustion. The pressure and temperature are set at 1 bar and 1400 °C (1673 K) in accordance with the work of Sivalingam [15]. The RGIBBS reactor automatically calculates the combustion products at equilibrium such that the Gibbs free energy is minimized. The airflow rate into the RGIBBS reactor is set at the minimum value where the carbon is completely combusted. If it is less, energy will remain trapped in the coal; if it is more, energy

Table 3.19 Physical and chemical properties of Illinois #6 coal

Proximate analysis (wt. %)				Ultimate analysis (wt. %)					
Moisture	Volatile matter	Fixed carbon	Ash	C	H	N	Cl	S	O
11.12	34.99	44.19	9.70	80.51	5.68	1.58	0.37	3.17	8.69

will be wasted in heating the excess air to the reactor temperature. The calculation for the proper amount of air is discussed in Sect. 3.5.1.2. After combustion, the materials are taken to a separator to isolate the ash from the other conventional materials. At this stage, the combustion process is considered complete; the CO_2-rich flue gas then undergoes the calcium looping process. The temperature of the flue stream is maintained at 150 °C in accordance with the lower limits on power plant flue gas temperatures provided by Feron [16].

2. Carbonator

The carbonator refers to the reactor where the carbonation reaction takes place. The RSTOIC reactor block is used in Aspen Plus to model the carbonator. The pressure is set at 1 bar and the temperature is set at 650 °C. The RSTOIC is a reactor in which the user can define the specific reaction that occurs. The carbonation reaction is given by

$$CaO\ (s) + CO_2\ (g) \rightarrow CaCO_3(s) \tag{3.5}$$

In real situations, CaO and CO_2 do not react completely with each other. The amount of CaO that can actually react is constrained by the surface area of CaO particles. Furthermore, the mixing between CO_2 and CaO is affected by how the fluidization develops in the reactor. These effects can be incorporated into Aspen Plus as the extent of the reaction by defining the conversion fraction for one of the reactants, CaO. The dependence of the CO_2 capture efficiency of the carbonator on the sorbent flow ratios is shown in Fig. 3.32 [15]. Figure 3.32 has been translated into a table format by Sivalingam [15]; a part of his table used in this paper is shown in Table 3.20. In Fig. 3.32 and Table 3.20, F_{CO_2} is the mole flow rate of CO_2, F_R is the mole flow rate of recycled (or looped) CaO, and F_0 refers to the make-up flow of CaO.

It is not possible to model the make-up flow of CaO in Aspen Plus. Hence, one value of F_0/F_{CO_2} is chosen to obtain one set of data for calculation. $F_0/F_{CO_2} = 0.1$ is chosen with three values of F_R/F_{CO_2} such that three values of CO_2 capture efficiency are modeled in the range of 50% to 100%. For a certain flow ratio and CO_2 capture efficiency, there is certain associated CaO conversion fraction. Since the CO_2 capture efficiency cannot be directly controlled, multiple cases are run in Aspen Plus for a certain CaO conversion fraction until the correct CO_2 capture efficiency is obtained. As shown in Table 3.21, each specified CaO conversion fraction corresponds to a range of CO_2 capture efficiencies. The simulation results for various CaO conversion fractions are plotted in Fig. 3.33. The small symbols (circles and triangles) in Fig. 3.33 represent all the trial cases conducted in Aspen Plus; these are called the results calculated from extrapolated data in future discussions. The large symbols refer to the cases whose results fit the data of Sivalingam [15]; these cases are called results obtained from experimental data in the future discussions.

Downstream of carbonator, the mixture of solids (primarily $CaCO_3$ with some CaO depending on the inlet flow rate of CaO) and the CO_2-lean flue gas is cooled

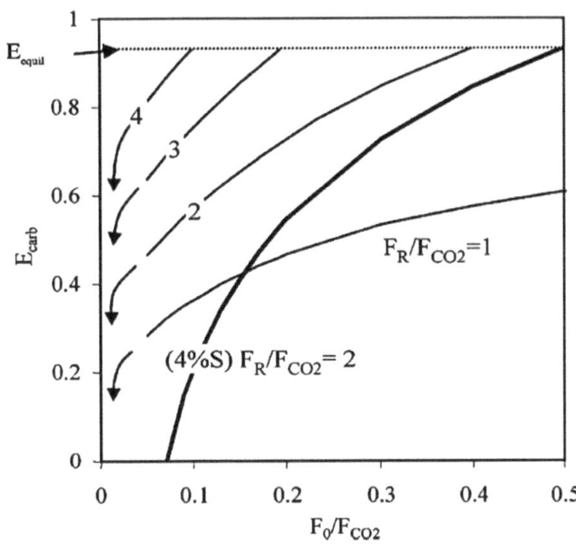

Fig. 3.32 CO_2 capture efficiency for different flow ratios of CaO and CO_2 [15]

Table 3.20 CO_2 capture efficiency for different flow ratios of CaO and CO_2 [15]

F_0/F_{CO_2}	$F_R/F_{CO_2} = 3$	$F_R/F_{CO_2} = 4$	$F_R/F_{CO_2} = 5$
0.05	0.63	0.81	0.99
0.10	0.76	0.95	0.99

Table 3.21 Range of CO_2 capture efficiencies for each CaO conversion fraction

F_R/F_{CO_2}	CaO conversion fraction	CO_2 capture efficiency
3	0.33	0.66–0.86
4	0.25	0.86–0.97
5	0.20	0.97–0.99

back to 150 °C by a heat exchanger and returns the heat released during the cooling-down process back to the carbonator. This step is necessary to account for the heat of the carbonator and the calciner separately. The stream then goes through a separator to separate the CO_2-lean flue gas from the solids, which are sent to the calciner to regenerate the CaO.

3. Calciner

Similar to the carbonator, the RSTOIC reactor block is employed for calciner in ASPEN Plus. The calcination reaction that takes place in the calciner is given by

$$CaCO_{3(s)} \rightarrow CaO_{(s)} + CO_{2(g)} \tag{3.6}$$

The temperature in this block is 900 °C, and the pressure is 1 bar in accordance with Sivalingam [15]. Unlike carbonation, the calcination reaction is a complete

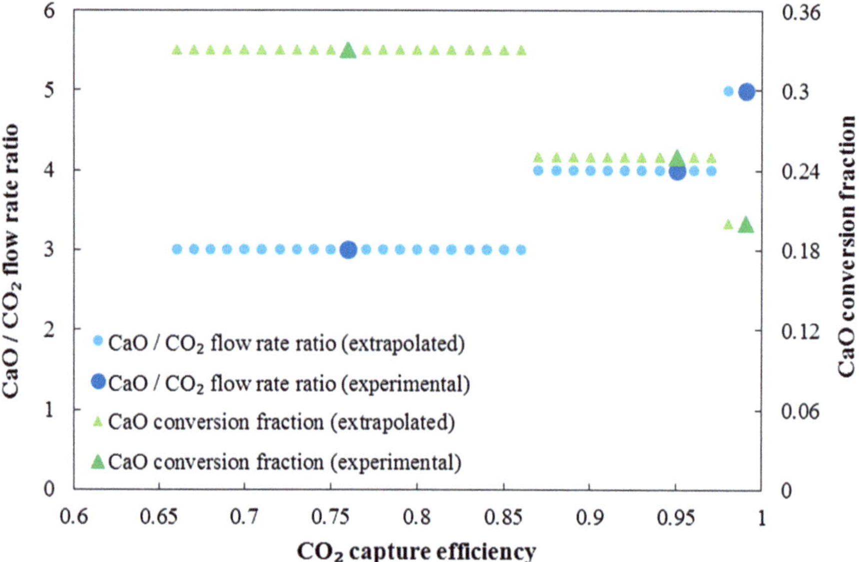

Fig. 3.33 Range of CO_2 capture efficiencies for various CaO conversion fractions

reaction, so all the $CaCO_3$ is decomposed. Therefore, the conversion fraction of $CaCO_3$ is set at 1. Post-stream of the calciner, the flow again passes through a heat exchanger and a separator like before. The heat absorbed from this heat exchanger is added to the heat of calciner.

4. Summary of post-combustion setup

The reactor blocks used in Aspen Plus for CaL with post-combustion capture are listed in Table 3.22 along with their functions and reactions.

For the heat stream, the heat is added from the decomposer, burner, heat exchanger for ash, and heat exchanger for flue gas together to be the heat of the coal combustion process without CaL. The rest of the heat from the carbonator, post-carbonator heat exchanger, calciner, and post-calciner heat exchanger is the heat of the calcium looping process. These values of heat without and with CaL and the CO_2 fraction in the final outlet flow are indicative of the performance of the CaL system with post-combustion capture. The heat values of carbonation are also of interest in the evaluation of the performance of calcium looping with post-combustion capture. Figure 3.34 shows the final flow sheet in Aspen Plus for CaL with post-combustion capture.

Table 3.22 Process models used for calcium looping with post-combustion capture setup in Aspen Plus

Name	Reactor model	Function	Reaction formula
DECOMP	RYIELD	Converts nonconventional into conventional	Coal $\rightarrow$ char + simple substances
BURN	RGIBBS	Burns coal with air	Char + simple substances + $O_2 \rightarrow CO_2 + H_2O$
CARBONATOR	RSTOIC	Carbonation	$CaO + CO_2 \rightarrow CaCO_3$
CALCINER	RSTOIC	Calcination	$CaCO_3 \rightarrow CaO + CO_2$
SEP-ASH	SSPLIT	Flue gas and ash separation	–
SEP-CAR	SEP	Flue gas (CO_2-lean) and Ca-solids separation	–
SEP-CAL	SEP	CO_2 and Ca-solids separation	–
COOL-A	HEATER	Ash cooler	–
COOL-B	HEATER	Flue gas cooler	–
COOL-C	HEATER	Cooler downstream of carbonator	–
COOL-D	HEATER	Cooler downstream of calciner	–

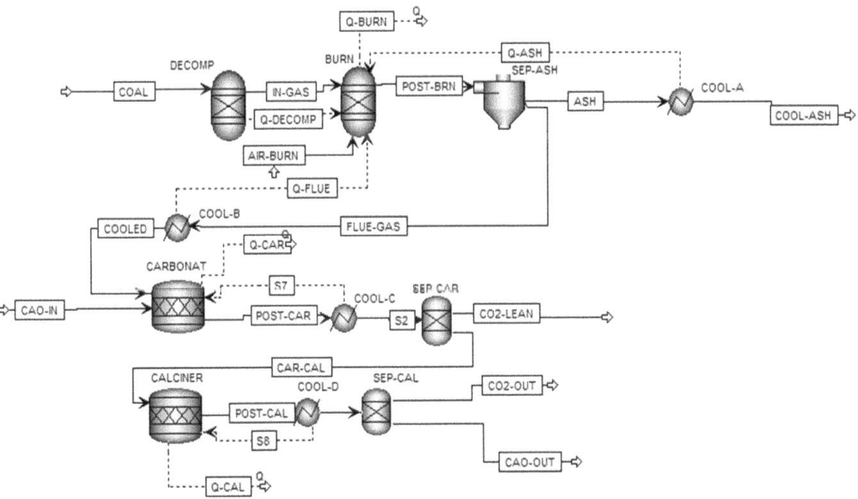

Fig. 3.34 Aspen Plus flow sheet for calcium looping with post-combustion capture

3.6.1.2   Results and Discussion

1. Calculation of the optimum airflow

It is not possible to use air as an input material in Aspen Plus. Instead, air is simulated as a mixture of approximately 21% O_2 and 79% N_2. The other components of air such as argon and CO_2 are present in such small fractions that their effect on the results is negligible. The optimization module in Aspen Plus is employed to find the flow rates of O_2 and N_2 such that the burner heat is maximized. The optimization setup and results are shown in Table 3.23 and Table 3.24, respectively. F_{CO}, F_{CO_2}, F_{O_2}, and F_{N_2} refer to the mole flow rate of various gases (CO and CO_2 downstream of the burner, O_2 and N_2 upstream of the burner).

The basic idea behind the optimization is to change O_2 and N_2 flow rates separately within a small range in order to get the maximum CO_2 flow out of the burner. Constraints for this process cover the flow rate of CO out of burner and ratio between O_2 and N_2. This ensures that nearly all the combustion product is CO_2 (indicating complete combustion) with the smallest amount of O_2 and N_2, while the ratio between O_2 and N_2 is the same as that in the air. The optimized flow rates of O_2 and N_2 are 4.20 kmol/s and 15.76 kmol/s, respectively, for 50 kg/s of coal. The sum of the flow rates of nearly 20 kmol/s indicates the total airflow in the burner. For scaled higher coal flow rate cases, the O_2 and N_2 flow rates are changed proportionally.

2. Carbonation and calcination analysis

Since the main objective of CaL is to reduce the amount of CO_2 released into the atmosphere, the CO_2 capture efficiency is the most important metric in evaluating the performance of this process. Therefore, the CO_2 capture efficiency used as an independent variable on the *x*-axis and all other quantities of interest are plotted on the *y*-axis to examine how the CO_2 capture efficiency affects the other quantities. Since there are no other energy losses that need to be considered, the energy penalty of CaL is the sum of the heat gain and loss from the carbonation and calcination.

For each CaO conversion fraction, there is a corresponding CO_2 capture efficiency from the experimental data [15]. Once the CaO conversion fraction is defined in Aspen Plus, the CO_2 capture efficiency is manipulated to be equal to the experimental result by changing the CaO inflow to the carbonator. With change in CaO inflow, the heat duty of the carbonator and calciner change. Therefore, for each set of experimental data, one can calculate one data point of heat duty for the carbonator and calciner. In order to obtain additional data points for the heat duty, the extrapolated data in Fig. 3.35 are considered as well, corresponding to a range of

Table 3.23 Optimization variables setup and results

	O_2	N_2
Manipulation range (kmol/s)	3–5	12–16
Maximum step size (kmol/s)	0.5	0.5
Optimized result (kmol/s)	4.20	15.76

Table 3.24 Constraints and convergence results for optimization of amount of air

Convergence criteria	Tolerance	Result
$F_{CO} = 0.001$ kmol/s	0.001 kmol/s	0.00059 kmol/s
$F_{O_2}/F_{N_2} = 0.2658$	0.01	0.2664
Maximized F_{CO_2}	–	2.653 kmol/s

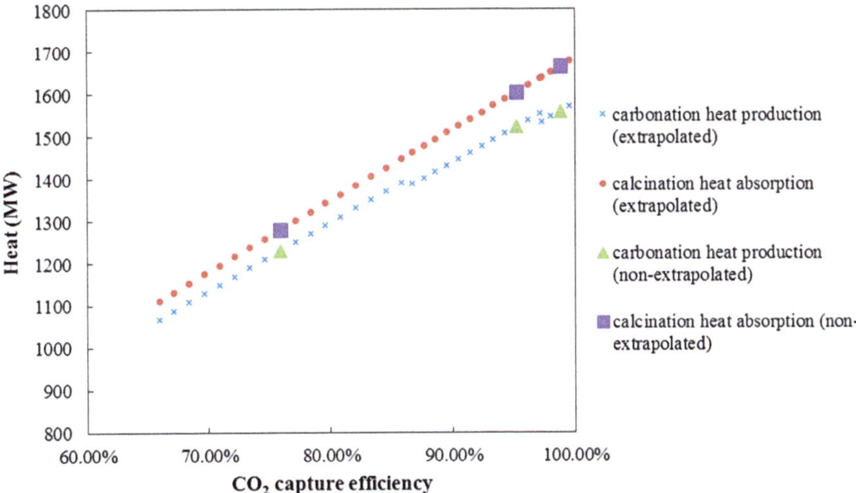

Fig. 3.35 Heat duty of carbonator and calciner for original experimental data and extrapolated data

CO_2 capture efficiencies for each CaO conversion fraction due to varying CaO flow rates instead of just the one that matches the result of Sivalingam [15]. The heat duty of the carbonator and calciner for both the results from experimental data (non-extrapolated) and extrapolated data are shown in Fig. 3.35.

It should be noted that the calcination reaction is an endothermic reaction since heat must be added to precipitate the decomposition of $CaCO_3$. Thus, the heat duty of the calciner from Aspen Plus is negative. However, the absolute value of the calciner heat is plotted in Fig. 3.35 to compare it with the heat production in the carbonator. For each capture efficiency, the heat absorbed by the calciner is greater than the heat produced in the carbonator, which confirms that there is a net energy penalty associated with the calcium looping process.

Extrapolating the data to account for additional CO_2 capture efficiencies is necessary because it is not possible to get an accurate representation of a trend from the only three experimental data points available. It can be observed from Fig. 3.35 that the calcination results form an almost perfect line and the extrapolated and non-extrapolated data are coincident with the line. This linearity is expected since the calculation is based on a stoichiometric relation, resulting in the heat produced also proportional to the inflow rate of the reactant, $CaCO_3$. The CaO conversion fraction does not affect the heat absorbed by the calciner. This is because the calciner has the same temperature for both inlet and outlet flow (150 °C). When there is an excess amount of CaO fed into the carbonator, the unreacted CaO will pass through the carbonator and enter the calciner; this unreacted part of CaO has no effect on the reaction within the calciner. Thus the heat duty of calciner remains unchanged by the excess amount of CaO.

For the carbonator, the extrapolated data around each of the three experimental data points are linear, but these lines do not coincide. The linearity of each section can be explained the same way as for the calciner. However, in this case, the CaO conversion fraction has some effect. Each straight line section corresponding to a range of extrapolated data has a reduced y-intercept compared to the previous section and has a more gradual slope compared to the calciner. From the modeling point of view, the only difference between these two reactors, besides the chemical reaction, is the inlet stream temperature. Since the CaO stream entering the carbonator is an external input, the inlet stream has a temperature of 25 °C compared to 150 °C for the calciner (internal input from the carbonator outlet stream). Thus, some heat is consumed in the carbonator for heating up the inlet stream to the temperature of the outlet. The heat production of the carbonator decreases as the CO_2 capture efficiency is increased since more heat is consumed to heat the higher CaO flow that is required for the increased capture. This is the main reason for the difference in behavior of the calciner and carbonator described above.

3. Energy penalty analysis

For 50 kg/s of inlet coal flow, the heat of combustion is calculated to be 1168 MW (i.e., without calcium looping). When the net heat gained and lost in the carbonator and calciner, respectively, are added, the total heat of the power plant with calcium looping ranges from 1060 to 1130 MW, as shown in Fig. 3.36. As expected based on the trends in calciner and carbonator heat, the total heat output of the power plant decreases as more carbon is captured.

The energy penalty for CaL refers to the fraction of energy produced by a power station that must be dedicated to the carbonation and calcination process in order to capture CO_2. The energy penalty can be defined as

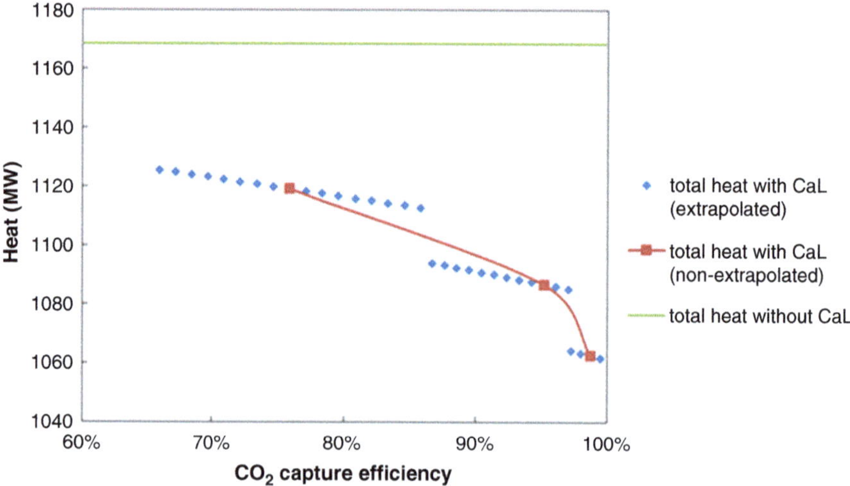

Fig. 3.36 Plot of total heat output vs. CO_2 capture efficiency without CaL and with post-combustion CaL

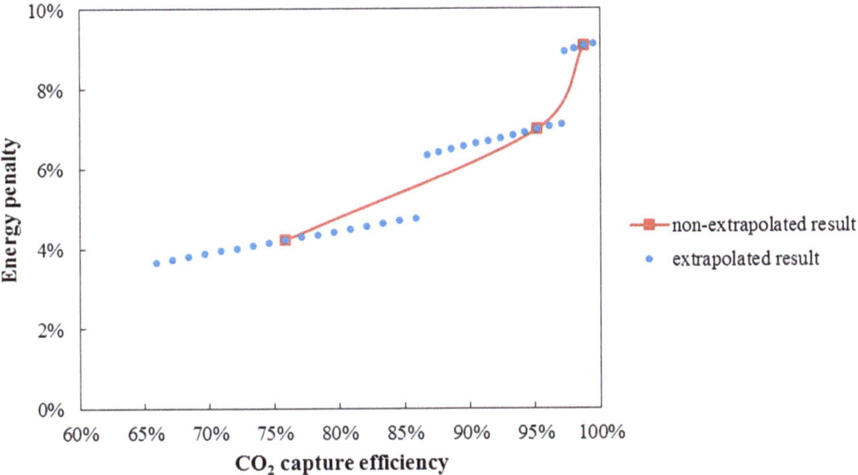

Fig. 3.37 Plot of energy penalty vs. CO$_2$ capture efficiency for CaL with post-combustion capture

$$\text{Energy penalty} = \frac{|Q_{\text{looping}}|}{Q_{\text{total}} - |Q_{\text{looping}}|} \tag{3.7}$$

where Q_{total} is the total heat produced by the power plant and Q_{looping} is the heat for the CaL process.

Figure 3.37 shows the energy penalty for CaL with post-combustion capture calculated using Eq. (3.7). From the figure, the energy penalty ranges from 3.5% to 9.0% over the corresponding range of CO$_2$ capture efficiencies from 65% to 99%.

Figure 3.37 follows the curve for the total energy (heat) shown in Fig. 3.36. From Figs. 3.36 and 3.37, it can be noted that the results obtained from the extrapolated data form three sections of straight lines with discontinuities. On the other hand, the curve fit for the non-extrapolated result connecting the three points from the experimental data also shows a gradually increasing slope. At first glance, the curve fit looks more reasonable, and the extrapolated data seems to be not an accurate representation because of the discontinuities. For more accurate calculations to determine how the energy penalty varies with the CO$_2$ capture efficiency between the experimental data points, the combination of these two methods may therefore be desirable. Since the blue dots show linear relationship in certain range, this information could be used to extrapolate between the blue dots for a given straight line. However, in the regions where there is jump between the two straight lines, it may be desirable to create an appropriate smooth curve.

It should be noted that only the heat output of the combustor and the CaL process is assessed in the results presented above. In a real plant, heat generated from coal combustion will heat up the steam to drive a steam turbine connected to an electric generator to generate electricity. Thus, it is the difference in temperatures that matters in the power plant. Even with some heat sources having the same amount of thermal energy, difference in temperatures can lead to different amounts of electricity generated by these sources. In many papers, the energy penalty is calculated based on the electric power generated by the whole power plant. However, since the steam cycle is not considered in this paper, the energy penalty is calculated based only on the heat of coal combustion; therefore the effect of high temperature is not reflected in the present model. It is assumed that all heat sources can contribute to energy until they reach a temperature as low as 150 °C. Furthermore, when considering the whole plant, transportation of the solid calcium will cost extra energy, although as discussed earlier this amount is likely to be negligible [17]. In summary, this paper considers a simplified model of calcium looping, which only takes into account the heat of chemical reaction in a stoichiometric fashion. As a result of the various simplifications, the energy penalty calculated in this paper should be considered as a lower bound for any investigation on calcium looping.

3.6.2   Calcium Looping with Pre-Combustion Capture

3.6.2.1   Process Simulation Setup

Calcium looping with pre-combustion capture is a technology that substantially alters the combustion process. Instead of burning the coal in a boiler, incomplete combustion takes place in a gasifier to convert the carbon into CO. Next, a water-gas-shift reaction occurs between the CO and injected steam (H_2O) to form H_2 and CO_2, which go into the carbonator and calciner loop to capture the CO_2 prior to the final step of combusting the H_2 for energy production.

1. Gasifier setup

   In a typical integrated gasification combined cycle power plant, oxygen from an air separation unit (ASU) is used to gasify the coal to form CO and H_2 while avoiding production of CO_2 and H_2O. However, the air separation unit has a significant high energy penalty associated with it. From an energy standpoint, as long as the initial reactants and the final products remain the same, the heat produced in the gasification process will remain the same. Thus, to simplify gasification and eliminate the energy penalty of ASU, steam gasification is considered in this section. Similar to the post-combustion setup described in the previous section, the RYIELD and RGIBBS reactor blocks are used in the pre-combustion setup as well. The difference is that the coal in the gasifier reacts with steam instead of air, unlike in the burner in the post-combustion setup. The pressure of the gasifier unit is set at 1 bar, and the temperature is set at 1400 °C, as before.

2. Carbonator and calciner

   The carbonator and calciner have the same setup as in the post-combustion capture since the calcium looping process runs independent of the combustion scenario. However, an additional RSTOIC reactor is added upstream of the carbonator to implement the water-gas-shift reaction. The stream of syngas out of the gasifier comprises of CO and H_2. The purpose of the shift reaction is to react the CO with steam to form CO_2 and H_2. These gases then undergo the CaL process to capture the CO_2 and isolate the H_2 for combustion to produce energy. The pressure and temperature in the shift reactor are set at 1 bar and 650 °C, respectively.

3. Hydrogen burner

In the post-combustion capture setup, the burning of coal in the combustor produces energy, part of which is absorbed in the calcium looping process. In pre-combustion capture, energy is produced by burning the H_2 in the CO_2-lean flue gas that comes out of the carbonator; both gasification and calcium looping processes absorb a part of this energy. Thus, in the pre-combustion setup in Aspen Plus, an additional RSTOIC reactor is used to burn the H_2 produced in air. The inlet air temperature for the hydrogen burner is set at 150 °C, which is the same as final outlet temperature. This is done in order to eliminate the influence of the airflow rate in the calculations such that a steady excess airflow value can be maintained. Otherwise, if the inlet air was at 25 °C and is in excess for the amount of H_2, some energy would be wasted in heating up the excess air to the outlet stream temperature of 150 °C.

4. Summary of pre-combustion model setup

The various reactor blocks used in ASPEN Plus and their specifications for pre-combustion capture are listed in Table 3.25. The flow sheet setup in ASPEN Plus is shown in Fig. 3.38.

3.6.2.2   Results and Discussion

1. Calculation of the optimum steam flow rate for gasifier

In the coal gasification step, the carbon content in the coal is turned into CO. The goal in this step is to determine the H_2O flow rate to maximize the CO flow into the water-gas-shift reactor. A sensitivity analysis is employed to achieve this goal. The steam flow rate is incrementally increased to 3 kmol/s, and the C, CO, and CO_2 outflow rates are monitored as shown in Fig. 3.39.

It can be observed from Fig. 3.39 that at a 2.1 kmol/s flow rate of H_2O, all the carbon content (C) in the coal is burnt and CO_2 starts to form, and the CO output begins to decrease. Thus, 2.1 kmol/s is the optimum H_2O flow rate for the gasifier.

2. Calculation of the optimum steam flow rate needed for shift reactor

The optimum steam flow rate for the water-gas-shift reactor is obtained in a similar manner. The outflow of CO, CO_2, and H_2 are monitored as the inlet flow rate of H_2O is increased to 3 kmol/s. The results, shown in Fig. 3.40, indicate that the

Table 3.25 Process models used for calcium looping with pre-combustion capture setup in Aspen Plus

Name	Reactor model	Function	Reaction formula
DECOMP	RYIELD	Converts nonconventional into conventional	Coal $\rightarrow$ char + simple substances
GASIFIER	RGIBBS	Gasifies coal with steam	Char + simple substances + $H_2O \rightarrow CO + H_2$ + volatile matter
CARBONAT1	RSTOIC	Water-gas-shift reaction	CO + volatile matter + $H_2O \rightarrow CO_2 + H_2$
CARBONAT2	RSTOIC	Carbonation	$CaO + CO_2 \rightarrow CaCO_3$
H2-BURN	RSTOIC	Burns H_2 in air	$H_2 + O_2 \rightarrow H_2O$
CALCINER	RSTOIC	Calcination	$CaCO_3 \rightarrow CaO + CO_2$
SEP-ASH	SSPLIT	Flue gas and ash separation	–
SEP-CAR	SEP	Flue gas (H_2) and Ca-solids separation	–
SEP-CAL	SEP	CO_2 and Ca-solids separation	–
COOL-A	HEATER	Ash cooler	–
COOL-B	HEATER	Flue gas cooler	–
COOL-C	HEATER	Cooler downstream of carbonator	–
COOL-D	HEATER	Cooler downstream of H_2 burner	–
COOL-E	HEATER	Cooler downstream of calciner	–

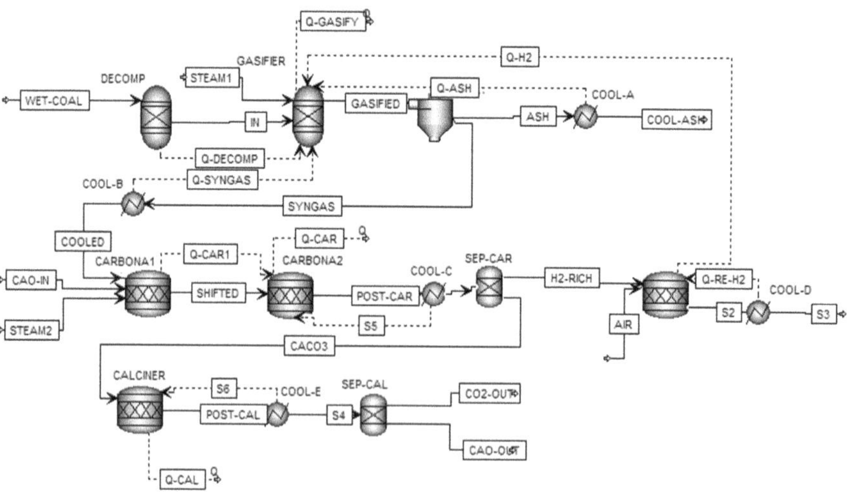

Fig. 3.38 Aspen Plus flow sheet for calcium looping with pre-combustion capture

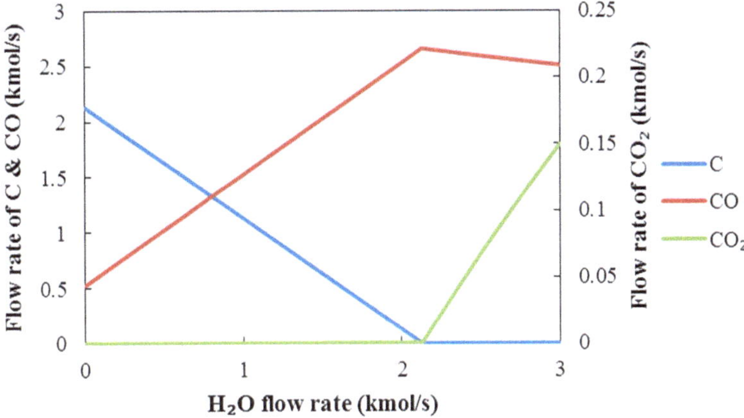

Fig. 3.39 Variation in gasifier outflow components with steam flow rate

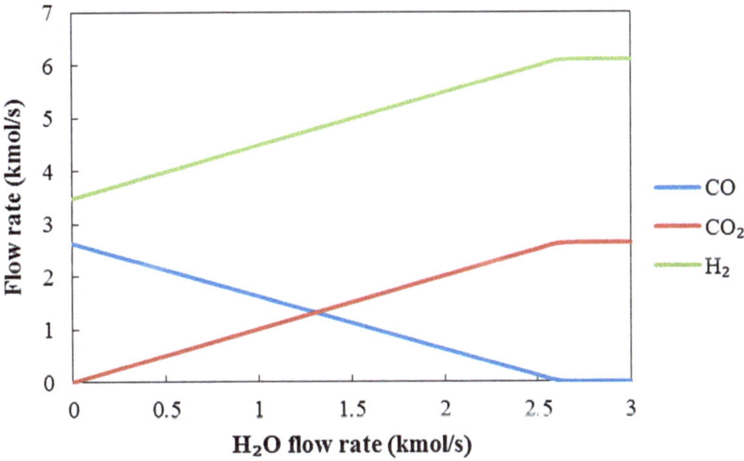

Fig. 3.40 Variation in shift reactor outflow components with steam flow rate

optimum steam flow in this case is 2.7 kmol/s when all the CO is converted and the CO_2 and H_2 flow rates acquire their highest values.

3. Energy penalty analysis

For pre-combustion capture without CaL, the net heat from the coal gasification and H_2 combustion combined is 1132 MW for a coal feeding rate of 50 kg/s. Taking into account the heat produced and absorbed in the carbonator and calciner, respectively, the total heat with CaL ranges from 980 to 1060 MW as shown in Fig. 3.41.

It should be noted that Fig. 3.41 shows a slightly lower total heat output without CaL compared to the post-combustion case because more energy is lost in

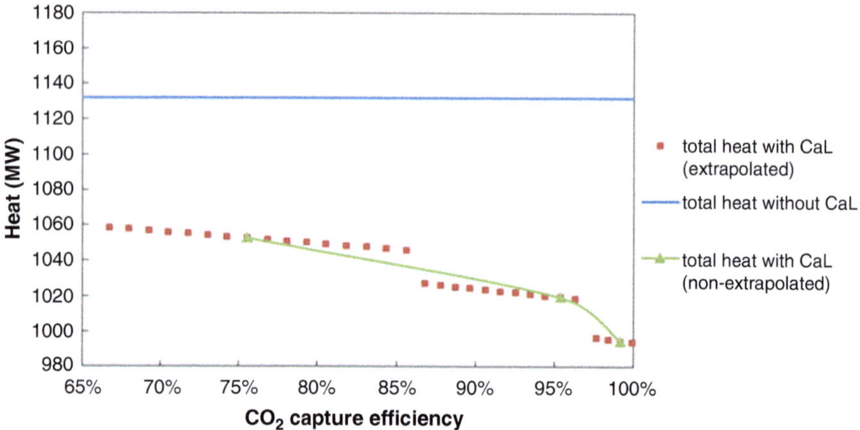

Fig. 3.41 Plot of total heat output vs. CO_2 capture efficiency without CaL and with pre-combustion CaL

pre-combustion from heating both steam and air from the external input temperature of 25 °C to the stream temperature of 150 °C. The various trends observed in Fig. 3.41 are similar to those observed in Fig. 3.36 for the post-combustion capture case.

Based on the results of Fig. 3.41, Eq. (3.7) is used to compute the energy penalty for different values of the CO_2 capture efficiency. The energy penalty for pre-combustion is compared to that of the post-combustion case in Fig. 3.42. It can be observed that the pre-combustion capture has a consistently higher energy penalty.

3.6.3 Scaling Considerations

The coal feeding rate considered in the simulations thus far is 50 kg/s following the work of Sivalingam [15]. In order to investigate the effect of a change of scale, CaL process models are developed with a smaller coal feeding rate of 5 kg/s with both post- and pre-combustion capture. For the downscaled case, the heat outputs for the post-combustion and pre-combustion cases are shown in Fig. 3.43. The total heat output is reduced by approximately an order of magnitude, in line with the order of magnitude reduction in the coal feeding rate. Even in the downscaled model, the total energy output in the pre-combustion case is smaller than that for the post-combustion case, as expected. Figure 3.44 shows the comparison between the energy penalty in pre-combustion and post-combustion case for the downscaled coal feeding rates. By overlaying the data for the original model from Fig. 3.42 onto Fig. 3.44, it is clear that the energy penalty is unaffected by the change in scale.

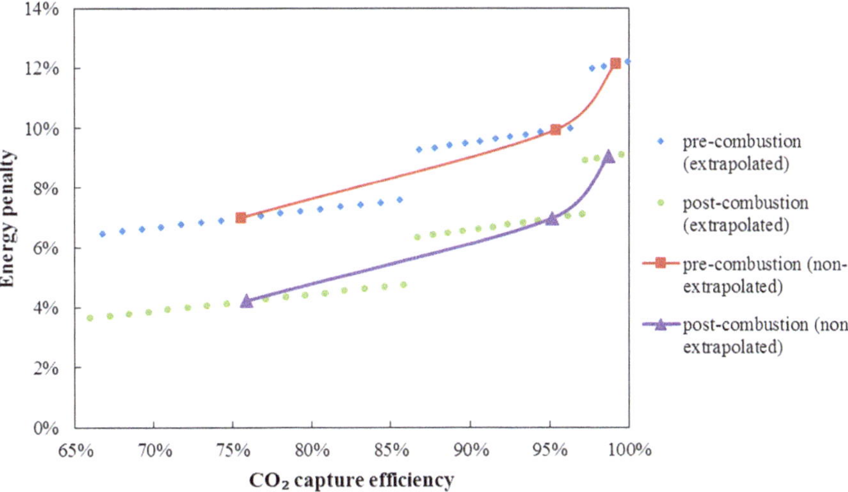

Fig. 3.42 Comparison of energy penalty vs. CO_2 capture efficiency for CaL with post- and pre-combustion capture

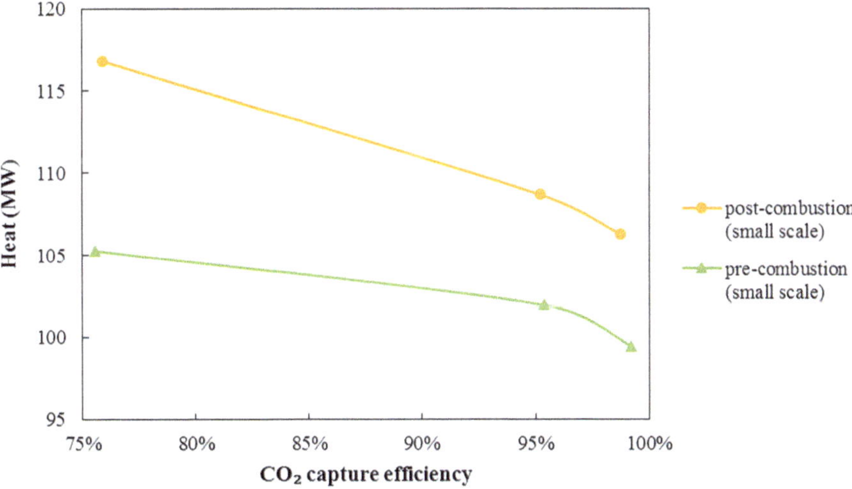

Fig. 3.43 Comparison of total heat output vs. CO_2 capture efficiency for CaL with post- and pre-combustion capture for the downscaled model with reduced coal feeding rate of 5 kg/s

3.6.4   Techno-Economic Analysis

After modeling and analyzing the two pre- and post-combustion calcium looping systems, a techno-economic analysis of the systems can be performed using Aspen Plus's Economic Analyzer. To do this, raw materials and stream prices are the input

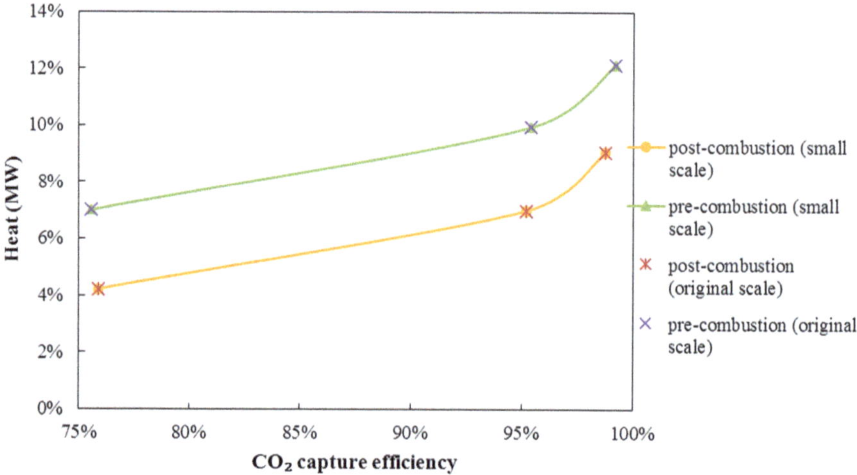

Fig. 3.44 Comparison of energy penalty vs. CO_2 capture efficiency for CaL with post- and pre-combustion capture for the original scale model and the downscaled model

Table 3.26 Input of raw materials and stream prices

Stream ID	Basis	Price	Unit
CaO-In	Mass	130	$/ton
CaO-Out	Mass	−126	$/ton
Wet-Coal	Mass	0.06	$/kg
Air	Mass	0	0

to the Economic Analyzer. The raw materials calcium oxide, coal, and air are input into the system. The rates at which the raw materials are input into the streams are adjusted in order to output 100MWh of energy. Table 3.26 shows the input of the raw materials and stream prices for the two calcium looping systems.

3.6.4.1   Results of Techno-Economic Analysis

1. Economic analysis of post-combustion calcium looping

The Aspen Plus model for post-combustion calcium looping is shown in Fig. 3.34. As shown in Table 3.27, the capital cost output of post-combustion calcium looping is $3,130,000, the raw material cost is $602,000,000, the operating cost is $667,000,000, and the utility cost is $14,500,000 for a period of 20 years. The total project capital cost breakdown is shown in Table 3.28. The total raw materials cost breakdown is shown in Table 3.29. The total operating cost breakdown is shown in Table 3.30. The total utility source cost breakdown is shown in Table 3.31. Note that in these Tables, one period is equal to 20 years.

Table 3.27 Economic summary of post-combustion calcium looping

Economic summary		
Total project capital cost	**3.13E + 06**	**Cost**
Total raw material cost	6.02E + 8	Cost/period
Total product sale	−4.91E + 08	Cost/period
Labor and maintenance cost	496,156	Cost/period
Total utility cost	1.45E + 07	Cost/period
Total operating cost	**6.67E + 08**	**Cost/period**
Operating labor costs	482,130	Cost/period
Maintenance cost	14,026	Cost/period
Operating charges	120,533	Cost/period
Plant overhead	248,078	Cost/period
Subtotal operating cost	6.17E + 08	Cost/period
G and A cost	4.94E + 07	Cost/period

Table 3.28 Total project capital cost breakdown of post-combustion calcium looping

Total project capital cost breakdown		
Purchased equipment	587,600	Total cost
Equipment setting	1530	Total cost
Piping	277,251	Total cost
Civil	16,943	Total cost
Steel	0	Total cost
Instrumentation	290,424	Total cost
Electrical	498,881	Total cost
Insulation	26,847	Total cost
Paint	0	Total cost
Other	932,700	Total cost
Subcontracts	0	Total cost
G and A overheads	62,483	Total cost
Contract fee	131,777	Total cost
Escalation	0	
Contingencies	508,758	Total cost
Total project cost	3.34E + 06	Total cost
Adjusted total project cost	3.13E + 06	Total cost

Table 3.29 Total raw material cost breakdown of - post-combustion calcium looping

Total raw materials cost breakdown		
Air-burn	0	Cost/h
CaO-In	57,848	Cost/h
Wet-coal	10,799	Cost/h

2. Economic analysis of pre-combustion calcium looping

The Aspen Plus model for pre-combustion calcium looping is shown in Fig. 3.38. The model is based on the post-combustion calcium looping model shown in Fig. 3.34. The difference is that instead of using the flue gases generated by the

Table 3.30 Total operating cost breakdown of post-combustion calcium looping

Total operating cost breakdown		
Operators per shift	1	
Unit cost	20	Cost/operator/h
Total operating labor cost	175,320	Cost/period
Maintenance cost/8000 h	12,800	Cost/period
Total maintenance cost	14,026	Total cost
Supervisors per shift	1	
Unit cost	35	Cost/supervisor/H
Total supervision cost	306,810	Cost/period

Table 3.31 Total utility cost breakdown of post-combustion calcium looping

Total utility cost breakdown						
Electricity		52.32	KW	KW	4.1	USD/h
Cooling water	Water	13.8	MMGAL	MMGAL/h	1653	USD/h

Table 3.32 Economic summary of pre-combustion calcium looping

Economic summary		
Total project capital cost	**8.45E + 06**	**Cost**
Total raw material cost	8.55E + 08	Cost/period
Total product sales	−7.43E + 08	Cost/period
Labor and maintenance	558,394	Cost/period
Total utility cost	1.16E + 07	Cost/period
Total operating cost	**9.37E + 08**	**Cost/period**
Operating labor costs	482,130	Cost/period
Maintenance cost	76,264	Cost/period
Operating charges	120,533	Cost/period
Plant overhead	279,197	Cost/period
Subtotal operating cost	8.68E + 08	Cost/period
G and A cost	6.94E + 07	Cost/period

complete combustion of coal to enter the carbonator and calciner, pre-combustion calcium looping uses incomplete combustion in a gasifier to convert the carbon into CO. In this gasifier, the coal reacts with steam instead of air in the post-combustion model. Next, a water-gas-shift reaction occurs between the CO and injected steam (H_2O) to form H_2 and CO_2, which enter the carbonator and the calciner loop to capture the CO_2 prior to the final step of combusting H_2 for energy production. An additional RSTOIC reactor is added upstream of the carbonator to implement this water-gas-shift reaction. Instead of burning coal to produce energy, pre-combustion calcium looping burns H_2 to produce energy. Thus, an additional RSTOIC reactor is used to burn the H_2 produced in air. As shown in Table 3.32, the capital cost output of post-combustion calcium looping is \$8,450,000, the raw material cost is \$855,000,000 per period, the operating cost is \$937,000,000 per period, and the utility cost is \$11,600,000 per period. The total project capital cost breakdown is shown in Table 3.33. The total raw materials cost breakdown is shown in Table 3.34.

Table 3.33 Total project capital cost breakdown of pre-combustion calcium looping

Total project capital cost breakdown		
Purchased equipment	2.17E + 06	Total cost
Equipment eetting	61,695	Total cost
Piping	774,977	Total cost
Civil	355,907	Total cost
Steel	116,227	Total cost
Instrumentation	439,616	Total cost
Electrical	507,719	Total cost
Insulation	236,886	Total cost
Paint	23,709	Total cost
Other	2.44E + 06	Total cost
Subcontracts	0	Total cost
G and A overheads	184,390	Total cost
Contract fee	323,494	Total cost
Escalation	0	Total cost
Contingencies	1.37E + 06	Total cost
Total project cost	9.01E + 06	Total cost
Adjusted total project cost	8.45E + 06	Total cost

Table 3.34 Total raw materials cost breakdown of pre-combustion calcium looping

Total raw materials cost breakdown		
Air-burn	0	Cost/h
CaO-In	86,772	Cost/h
Wet-coal	10,798	Cost/h

Table 3.35 Total operating cost breakdown of pre-combustion calcium looping

Total operating cost breakdown		
Operators per shift	1	
Unit cost	20	Cost/operator/h
Total operating labor cost	175,320	Cost/period
Maintenance cost/8000 h	69,600	Cost/period
Total maintenance cost	76,264	Total cost
Supervisors per shift	1	
Unit cost	35	Cost/supervisor/h
Total supervision cost	306,810	Cost/period

The total operating cost breakdown is shown in Table 3.35. The total utility source cost breakdown is shown in Table 3.36. Note that in these tables, one period is equal to 20 years.

3.6.4.2   Discussion

In this section, process models of calcium looping developed using Aspen Plus are used to determine the techno-economic cost of the CaL system by employing the

Table 3.36 Total utility cost breakdown of pre-combustion calcium looping

Total utility cost breakdown						
Electricity		52.32	KW	KW	0.000002	USD/h
Cooling water	Water	11.07463	MMGAL	MMGAL/H	1326.9958	USD/h

Aspen Economic Analyzer. The techno-economic analysis of the Aspen Plus models have shown that CaL is an economically viable system to implement in power plants, steel mills, and cement factories.

The results show that post-combustion calcium looping is associated with lower economic costs than that of the pre-combustion calcium looping. The total project capital cost of pre-combustion looping is \$5.32E + 06 higher than the total project cost of post-combustion looping. The association of pre-combustion calcium looping with the higher total project capital cost can be attributed to three reasons. The first reason is that in pre-combustion calcium looping, an additional cost of a RSTOIC reactor is incurred which needs to be added upstream of the carbonator to implement the water-gas-shift reaction before the stream of syngas coming out of the gasifier is reacted with steam to form CO_2 and H_2. The second reason is that in pre-combustion calcium looping, H_2 is burned to form energy. To do this, an additional cost of another RSTOIC reactor is incurred. Thus, in pre-combustion calcium looping, the need for additional equipment (two RSTOIC reactors) increases the total project capital cost. The third reason is the requirements of utility infrastructure. Because of the additional equipment needed in pre-combustion calcium looping, additional costs of equipment setting, piping, insulation, and others are also added. Thus, these three reasons attribute to higher associated total project cost of pre-combustion calcium looping compared to post-combustion calcium looping.

The results show that the total operating costs correlate to total project cost because pre-combustion calcium looping is associated with the higher cost. The total operating cost of pre-combustion looping is \$2.70E + 08 higher than the total operating cost of post-combustion looping. This can be attributed to the energy efficiency. Previous studies have shown that post-combustion looping is able to produce energy more efficiently than pre-combustion looping and thus requires less operating hours and supervision to produce the same amount of energy as pre-combustion looping.

Total project capital cost and total operating costs are the costs that determine the economic viability of the two CaL systems. Because the total project cost and the total operating cost of the post-combustion calcium looping system are lower than those of pre-combustion calcium looping, it can be concluded that post-combustion calcium looping is economically more viable system for carbon capture between the two CaL systems.

References

1. Aspen Plus®, 2019. Aspen Plus® V11.0 Documentation
2. Sahir A.H., A.L. Cadore, J.K. Dansie, N. Tingey, J. Lighty, Process analysis of chemical looping with oxygen uncoupling (CLOU) and chemical looping combustion (CLC) for solid fuels. In *Proceedings of the 2nd International Conference on Chemical Looping*, 2012
3. H. Leion, A. Lyngfelt, T. Mattisson, Solid fuels in chemical-looping combustion using a NiO-based oxygen carrier. Chem. Eng. Res. Des. **87**(11), 1543–1550 (2009)
4. T. Mattisson, A. Lyngfelt, H. Leion, Chemical-looping with oxygen uncoupling for combustion of solid fuels. Int. J. Greenh. Gas Control **3**(1), 11–19 (2009)
5. H. Leion, T. Mattisson, A. Lyngfelt, Using chemical-looping with oxygen uncoupling (CLOU) for combustion of six different solid fuels. Energy Procedia **1**(1), 447–453 (2009)
6. T. Mattisson, H. Leion, A. Lyngfelt, Chemical-looping with oxygen uncoupling using CuO/ZrO2 with petroleum coke. Fuel **88**(4), 683–690 (2009)
7. M. Rydén, A. Lyngfelt, T. Mattisson, Combined manganese/iron oxides as oxygen carrier for chemical looping combustion with oxygen uncoupling (CLOU) in a circulating fluidized bed reactor system. Energy Procedia **4**, 341–348 (2011)
8. A. Abad, I. Adánez-Rubio, P. Gayán, F. García-Labiano, F. Luis, J. Adánez, Demonstration of chemical-looping with oxygen uncoupling (CLOU) process in a 1.5 KWth continuously operating unit using a Cu-based oxygen-carrier. Int. J. Greenh. Gas Control **6**, 189–200 (2012)
9. L. Zhou, Z. Zhang, C. Chivetta, R. Agarwal, Process simulation and validation of chemical-looping with oxygen uncoupling (CLOU) process using Cu-based oxygen carrier. Energy Fuel **27**(11), 6906–6912 (2013)
10. X. Zhang, S. Banerjee, L. Zhou, R. Agarwal, Process simulation and maximization of energy output in chemical-looping combustion using ASPEN plus. Int. J. Energy Environ. **6**(2), 201 (2015)
11. T. Mendiara, A. Pérez-Astray, M.T. Izquierdo, A. Abad, L.F. De Diego, F. García-Labiano, P. Gayán, J. Adánez, Chemical looping combustion of different types of biomass in a 0.5 KWth unit. Fuel **211**, 868–875 (2018)
12. W.X. Meng, S. Banerjee, X. Zhang, R.K. Agarwal, Process simulation of multi-stage chemical-looping combustion using Aspen Plus. Energy **90**, 1869–1877 (2015)
13. L.H. Shen, J.H. Wu, J. Xiao, Q.L. Song, R. Xiao, Chemical-looping combustion of biomass in a 10 KW(Th) reactor with iron oxide as an oxygen carrier. Energy Fuel **23**(5–6), 2498–2505 (2009)
14. H.M. Gu, L.H. Shen, J. Xiao, S.W. Zhang, T. Song, Chemical looping combustion of biomass/coal with natural iron ore as oxygen carrier in a continuous reactor. Energy Fuel **25**(1), 446–455 (2011)
15. S. Sivalingam, *CO2 Separation by Calcium Looping from Full and Partial Fuel Oxidation Processes* (Technische Universität, München, 2013)
16. P. H. M. Feron, M. I. Attalla, G. Puxty, A. Allport, A. J. Cottrell, J. A. McGregor, Post-combustion capture (PCC) R&D and pilot plant operation in Australia. In *IEA GHG 11th Post Combustion CO2 Capture Meeting*, Vienna, Austria, 2008
17. A. Lyngfelt, B. Leckner, T. Mattisson, A fluidized-bed combustion process with inherent CO2 separation, application of chemical-looping combustion. Chem. Eng. Sci. **56**(10), 3101–3113 (2001)

Chapter 4
Computational Fluid Dynamics Modeling Methodologies

4.1 Two-fluid Model

The Eulerian or multi-fluid model circumvents the high computational demand of the particle-based models. In the multi-fluid models, the solid phase is also considered as a continuum fluid, and particle variables such as mass, velocity, temperature, etc. are averaged over a region that is large compared to the particle size. As such, this approach only accounts for the bulk behavior of the solid phase. Constitutive equations for the solid phase pressure and solid phase viscosity, which are required to model the interactions between the solid and the gas phases, are provided by the kinetic theory of granular flow. The kinetic theory of granular flow is an extension of the classical kinetic gas theory that includes the inelastic particle-particle interactions. The multi-fluid models are a simpler approach to investigate the fluid properties compared to the particle-based method and have been successfully implemented in a wide range of fluidization applications such as bubbling fluidized beds, particle mixing, down-flow reactors, and spouted beds. Jung and Gamwo [1] were the first to apply the multi-fluid approach for modeling the CLC process. However, the drawback of the Eulerian multi-fluid model is that the exact particle dynamics in the system cannot be determined and optimized.

For multiphase simulations using the Eulerian approach, the standard equations of fluid motion are slightly modified to account for the presence of additional phases by including the variable α defined as the volume fraction of the respective phase in the computational cell where the hydrodynamic equations are applied. The continuity equation for phase q is given as

$$\frac{\partial}{\partial t}(\alpha_q \rho_q) + \nabla \cdot (\alpha_q \rho_q \boldsymbol{u}_q) = \sum (\dot{m}_{pq} - \dot{m}_{qp}) \tag{4.1}$$

R. K. Agarwal, Y. Shao, *Modeling and Simulation of Fluidized Bed Reactors for Chemical Looping Combustion*, https://doi.org/10.1007/978-3-031-11335-2_4

where $\dot{m}_{pq}$ is the mass transfer rate from the pth phase to the qth phase. Each phase (gas or solid) consists of a number of species. A transport equation is solved for each species given as

$$\frac{\partial}{\partial t}\left(\alpha_q \rho_q Y_{iq}\right) + \nabla \cdot \left(\alpha_q \rho_q \boldsymbol{u}_q Y_{iq}\right) = \sum\left(\dot{m}_{ij}^{qp} - \dot{m}_{ji}^{pq}\right) \tag{4.2}$$

where Y_{iq} is the mass fraction of the species i in the qth phase and $\dot{m}_{ij}^{qp}$ is the mass transfer rate from the jth species of the pth phase to the ith species in the qth phase. Generally, one gas phase and one solid phase are considered corresponding to the fuel-gas mixture and the oxygen carrier, respectively.

The momentum equation for the gas phase is given as

$$\frac{\partial}{\partial t}\left(\alpha_g \rho_g \boldsymbol{u}_g\right) + \nabla \cdot \left(\alpha_g \rho_g \boldsymbol{u}_g \boldsymbol{u}_g\right) = -\alpha_g \nabla p + \nabla \cdot \bar{\bar{\tau}}_g + \alpha_g \rho_g \boldsymbol{g} + \sum\left(\boldsymbol{R}_{sg} + \dot{m}_{sg}\boldsymbol{u}_{sg} - \dot{m}_{gs}\boldsymbol{u}_{gs}\right) \tag{4.3}$$

where the terms in the summation are source terms added to the standard form of the Navier-Stokes momentum equations to account for the momentum transfer between the solid phase and the gas phase. Specifically, $\boldsymbol{R}_{sg} = \beta_{sg}(\boldsymbol{u}_s - \boldsymbol{u}_g)$ is the momentum transfer due to interphase drag, and the other terms are due to the transfer of mass. The momentum equation for the solid phase follows from the momentum equation for the gas phase with the source term for interphase drag being equal but opposite as given below:

$$\frac{\partial}{\partial t}\left(\alpha_s \rho_s \boldsymbol{u}_s\right) + \nabla \cdot \left(\alpha_s \rho_s \boldsymbol{u}_s \boldsymbol{u}_s\right) = -\alpha_s \nabla p + \nabla \cdot \bar{\bar{\tau}}_s + \alpha_s \rho_s \boldsymbol{g} + \sum\left(\boldsymbol{R}_{rs} + \dot{m}_{rs}\boldsymbol{u}_{rs} - \dot{m}_{sr}\boldsymbol{u}_{sr}\right) \tag{4.4}$$

For the flow conditions in a fuel reactor, the gas can be considered as an incompressible fluid. The fluid stress tensor is simply the Cauchy stress tensor with zero bulk viscosity given as

$$\bar{\bar{\tau}}_g = \alpha_g \mu_g \left(\nabla \boldsymbol{u}_g + \nabla \boldsymbol{u}_g^{\mathrm{T}}\right) \tag{4.5}$$

On the other hand, the granular solid stress tensor considers all terms in the Cauchy stress tensor as given below:

$$\bar{\bar{\tau}}_s = -p_s \bar{\bar{I}} + \alpha_s \mu_s \left(\nabla \boldsymbol{u}_s + \nabla \boldsymbol{u}_s^{\mathrm{T}}\right) + \alpha_s \lambda_s (\nabla \cdot \boldsymbol{u}_s)\bar{\bar{I}} \tag{4.6}$$

where p_s is the solids pressure, μ_s is the granular viscosity, and λ_s is the granular bulk viscosity. The definition of these terms and the interphase exchange coefficient β_{sg} provide the basis for the Eulerian approach for multiphase flow simulation. The solids pressure and granular bulk viscosity are defined according to Lun et al. [2]; the granular viscosity is according to Gidaspow [3].

The energy equation for phase q is expressed in terms of the enthalpy as

$$\frac{\partial}{\partial t}\left(\alpha_q \rho_q h_q\right) + \nabla \cdot \left(\alpha_q \boldsymbol{u}_q h_q\right) = \alpha_q \frac{\partial p}{\partial t} + \nabla \cdot \left(\bar{\bar{\tau}}_q \cdot \boldsymbol{u}_q\right) - \nabla \cdot \boldsymbol{q}_q + S_q + \sum \boldsymbol{Q}_{pq}$$

(4.7)

where h_q and $\boldsymbol{q}_q$ are the specific enthalpy and heat flux of phase q, respectively. As with the continuity and momentum equations, source terms are implemented to account for the transfer of enthalpy between the phases. In particular, S_q is the enthalpy source due to chemical reaction, and $\boldsymbol{Q}_{pq}$ is the heat transfer from the pth phase to the qth phase. The interphase heat transfer is modeled based on Gunn [4].

4.2   CFD-DEM Model

In the Lagrangian particle-based model or discrete element method (DEM), the trajectory of each individual particle is determined based on a force balance calculation for the particle. The particle tracking is coupled with CFD solution for the fluid phase by considering the interaction between the particles and the fluid separately for each particle. DEM can model the properties such as temperature, composition, position, and velocity with high accuracy, limited only by the specifics of the particle collision parameters employed in the model. Since the number of particle collisions increases drastically with an increase in the number of particles, the number of particles in a DEM simulation is constrained because of limitation of available computational capability; this number can be increased if greater computing power is available. The high computational cost is the reason behind the scarcity of particle-based models for CLC simulation in the literature to date. However, cold-flow simulations using the coupled CFD-DEM model have proven capable in accurately matching the particle dynamics of various laboratory-scale fluidized bed experiments using relatively large Geldart Group D particles. It is crucial to combine the solid particle dynamics and chemical reactions into one credible model for the whole CLC fuel reactor that can be used to investigate various aspects of solid-fueled CLC process including reactor design, inlet jet velocity, and the physical properties of the oxygen carrier in order to achieve an optimized design in the future.

The equations for mass and momentum conservation for the fluid phase are identical to those used in the Eulerian model given in Eqs. (4.2) and (4.3) with the exception that the source term in Eq. (4.3) for the solid-gas momentum exchange term, $\boldsymbol{R}_{sg}$, is obtained from the average of the drag forces acting on all the discrete particles in a given computational cell. The shear stress term in the momentum equation is given in Eq. (4.5). The equation for the conservation of energy for the fluid phase can be expressed in terms of the internal energy as

$$\frac{\partial}{\partial t}\left(\alpha_f \rho_f E_f\right) + \nabla \cdot \left(\alpha_f \boldsymbol{u}_f\left(\rho_f E_f + p_f\right)\right) = \nabla \cdot \left(k_f \nabla T_f - \sum h_j \boldsymbol{J}_j + \left(\bar{\bar{\tau}}_f \cdot \boldsymbol{u}_f\right)\right) + S_h$$

$$(4.8)$$

where ρ_f, $\boldsymbol{u}_f$, p_f, E, and T are the density, velocity, pressure, internal energy, and temperature of the fluid, respectively, $\boldsymbol{g}$ is the acceleration due to gravity, k is the thermal conductivity, and h_j and $\boldsymbol{J}_j$ are the enthalpy and diffusion flux of species j. The source term, S_h, in Eq. (4.8) captures the net heat flux due to heat transfer from the solid to the gas phase due to the chemical reactions. Similar to the source term for momentum, it is calculated based on the average of all the discrete particles in the computational cell where the equation is applied.

The trajectory of each particle is computed by integrating the force balance on the particle, which can be written in the Lagrangian frame per unit particle mass as

$$\frac{\partial \boldsymbol{u}_p}{\partial t} = \boldsymbol{g}\frac{\left(\rho_f - \rho_p\right)}{\rho_p} + F_D\left(\boldsymbol{u}_f - \boldsymbol{u}_p\right) + \boldsymbol{F}_{col} \qquad (4.9)$$

where the subscript p denotes an individual particle. The terms on the right-hand side of Eq. (4.9) account for the gravitational and buoyant forces, the drag force, and an additional force due to particle-particle or particle-wall collisions. Forces such as the virtual mass force and pressure gradient force can be neglected for gas-solid flows since ρ_p far exceeds ρ_f. The net drag coefficient F_D is given by

$$F_D = \frac{18\mu_f}{\rho_p d_p^2}\frac{C_D Re_p}{24} \qquad (4.10)$$

where d_p is particle diameter, C_D is the particle drag coefficient, and Re_p is the Reynolds number based on the particle diameter defined as

$$\mathrm{Re}_p = \frac{\rho_f d_p \left|\boldsymbol{u}_f - \boldsymbol{u}_p\right|}{\mu_f} \qquad (4.11)$$

Here, the collision force in Eq. (4.9) is computed using the soft-sphere model, which decouples its normal and tangential components. The normal force on a particle involved in a collision is given by

$$\boldsymbol{F}_{col}^n = \left(K\delta + \gamma(\boldsymbol{u}_{12}\boldsymbol{e})\right)\boldsymbol{e} \qquad (4.12)$$

In Eq. (4.12), δ is the overlap between the particles pair involved in the collision as illustrated in Fig. 4.1 and γ is the damping coefficient, a function of the particle coefficient of restitution η; $\boldsymbol{e}$ is the unit vector in the direction of $\boldsymbol{u}_{12}$. Previous research by Link has demonstrated that for large values of K, the results of the

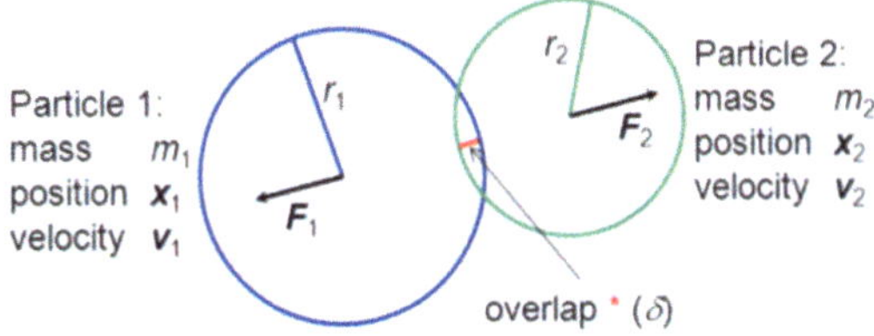

Fig. 4.1 Schematic of particle collision model for DEM

soft-sphere model are identical to those obtained using a hard-sphere model [5]. The tangential collision force is a fraction μ of the normal force with μ as a function of the relative tangential velocity v_r given as

$$\mu(v_r) = \begin{cases} \mu_{\text{stick}} + \left(\mu_{\text{stick}} - \mu_{\text{glide}}\right)\left(v_r/v_{\text{glide}} - 2\right)\left(v_r/v_{\text{glide}}\right) & \text{if } v_r < v_{\text{glide}} \\ \mu_{\text{glide}} & \text{if } v_r \geq v_{\text{glide}} \end{cases} \tag{4.13}$$

The chemical reactions in the flow require additional equations to compute the local mass fraction of each species Y_j in the computational cell. The species conservation equation is given by

$$\frac{\partial}{\partial t}\left(\rho Y_j\right) + \nabla \cdot \left(\rho \boldsymbol{u} Y_j\right) = -\nabla \cdot \boldsymbol{J}_j + R_j + S_j \tag{4.14}$$

where $\boldsymbol{J}_j$ is the diffusion flux of the species due to concentration gradients in the flow field, R_j is the net rate of production of the species due to chemical reactions, and S_j is the rate of creation of the species from devolatilization.

Heat transfer to the particle is governed by the equation for particle heat balance, which can be written as

$$m_p c_p \frac{dT_p}{d_t} = hA_p\left(T_\infty - T_p\right) - f_h \frac{dm_p}{dt} H_{\text{reac}} \tag{4.15}$$

where c_p is the particle heat capacity, h is the convective heat transfer coefficient, H_{reac} is the heat released by the reaction, and f_h is the fraction of the energy produced that is captured by the particle; the remaining portion $(1 - f_h)$ is applied as the heat source in the energy equation. Values of f_h can range from 1.0 for incomplete combustion where all the heat is retained by the particle (e.g., char combustion to form CO) to zero implying all the heat is released to the continuous phase [6]. Since the reduction of the metal oxide is considered to be a complete reaction, the default value of zero can be used.

4.3 Drag Models

In the gas-solid two-phase flow, the interactions between the gas phase and solid phase are very complex, which may result in the drag force, Brownian motion force, buoyancy force, virtual mass force, Bassett force, Magnus force, Saffman lift force, pressure gradient force, etc. depending upon the particle and fluid properties and nature of the flow field. The gas-solid drag force reflects the interaction and momentum exchange between gas and solid phases and determines the entrainment and transport process of solid particles. Compared to other forces, the drag force has dominant influence in the gas-solid flow characteristics. The accuracy of the drag force therefore has great influence in the overall prediction accuracy of the numerical simulation. The value of the drag force is closely related to several factors such as the turbulent motion of the gas phase, the relative velocity of the gas-solid two phases, the particle diameter, the particle concentration, etc.

Researchers have conducted a large number of experimental studies on the force acting on a single particle and found that the drag force is related to the gas-solid slip velocity and the cross-sectional area of the particles in the direction of the flow. However, the relationship is difficult to express as a unified expression for different ranges of Reynolds numbers. The drag coefficient of a single particle is expressed as

$$C_{Do} = \frac{\boldsymbol{F}_{d0}}{\frac{1}{2}\rho_g U_{\text{slip}}^2 A_{p0}} \tag{4.16}$$

where $\boldsymbol{F}_{d0}$ is the drag force acting on a single particle in a uniform gas flow, $\boldsymbol{U}_{\text{slip}}$ is the apparent gas-solid slip velocity, and A_{p0} is the projected area of a single particle in the direction of the incoming flow, and it is equal to $\pi d_p^2/4$ for a spherical particle. The drag force acting on a single particle is related to the Reynolds number $Re_0 = d_p \rho_g U_{\text{slip}}/\mu_g$.

In a fluidized bed, the drag force acting on a group of particles is different from that acting on a single particle. Many fluidization experiments have been conducted to establish the drag force models. The commonly used drag models for simulation of fluidized beds include Wen and Yu drag model, Ergun equation, Gidaspow drag model, Syamlal and O'Brien drag model, and energy minimization multi-scale model.

4.3.1 Wen and Yu Drag Model

Wen and Yu (1966) developed a drag model by deriving a correction to the single-particle drag coefficient. When the fluidized bed reaches the steady state, the model assumes that the drag force of a single particle $\boldsymbol{F}_{ds}$ is equal to its gravity. The ratio of $\boldsymbol{F}_{ds}$ to the drag force acting on a single particle in an infinite flow field $\boldsymbol{F}_{d0}$ is expressed as

$$\frac{F_{ds}}{F_{d0}} = \frac{-\frac{1}{6}\pi d_p^3(\rho_p - \rho_g)g}{C_{do}\frac{1}{2}\rho_g(u_g - u_p)^2 \frac{\pi d_p^2}{4}} = \frac{Ar}{18\,Re + 2.70Re^{1.687}} \tag{4.17}$$

where $Re = d_p\rho_g U_{\text{slip}}/\mu_g$ and $Ar = d_p^3\rho_g(\rho_p - \rho_g)g/\mu_g^2$.

From the experiments conducted for different types of fluid-solid systems, it was found the ratio of F_{ds} to F_{d0} can be written as a function of the voidage α_g:

$$\frac{F_{ds}}{F_{d0}} = f(\alpha_g) = \frac{C_D}{C_{D0}} = \alpha_g^{-4.65} \tag{4.18}$$

The drag force acting on a group of particles per unit volume is obtained as

$$F_d = \frac{1 - \alpha_g}{\frac{\pi}{6}d_p^3}\frac{1}{2}\rho_g(u_g - u_p)|u_g - u_p|\varepsilon_g^2\frac{\pi d_p^2}{4}C_{D0}\alpha_g^{-4.65} \tag{4.19}$$

Thus, the Wen and Yu drag coefficient can be written as

$$\beta_{sg} = \frac{3}{4}\frac{1 - \alpha_g}{d_p}\rho_g|u_g - u_p|C_{D0}\alpha_g^{-1.65} \tag{4.20}$$

4.3.2　Ergun Drag Model Equation

Ergun (1952) calculated the resistance loss between particles by measuring the static pressure drop when the air flow passed through a fixed bed. According to the Ergun equation, the pressure drop has the following expression for spherical particles:

$$\frac{\Delta P}{H} = 150\frac{\mu_g(1 - \alpha_g)^2 U_g}{\alpha_g^3 d_p^2} + 1.75\frac{(1 - \alpha_g)\rho_g U_g^2}{\alpha_g^3 d_p} \tag{4.21}$$

where ΔP is the pressure drop, H is the bed height, and U_g is the gas velocity in the empty bed. The first term on the right side of Eq. (4.21) is the pressure drop loss due to viscosity, and the second term is the pressure drop loss due to inertia. When the gas velocity in empty bed is replaced by apparent gas-solid slip velocity $U_g = \alpha_g\,|\,u_g - u_p|$, the balance between the drag force and pressure drop can be expressed as

$$\frac{\Delta P}{H}\alpha_g = \beta_A|u_g - u_p| \tag{4.22}$$

Thus, the drag coefficient proposed by Ergun is obtained as

$$\beta_{sg} = 150\frac{\left(1-\alpha_g\right)^2\mu_g}{\alpha_g d_p^2} + 1.75\frac{\left(1-\alpha_g\right)\rho_g\left|\boldsymbol{u}_g - \boldsymbol{u}_p\right|}{d_p} \tag{4.23}$$

4.3.3 Gidaspow Drag Model

The Gidaspow model is well suited for fluidized bed simulations that include a range of solid loadings from dilute to densely packed because it accounts for the differences in the solid-gas interaction behavior in the dilute and dense regions by switching between the drag prediction of the Ergun equation and the drag model of Wen and Yu based on the solids fraction α_s.

For $\alpha_g > 0.8$, the Gidaspow model for the exchange coefficient β_{sg} is obtained as

$$\beta_{sg} = \frac{3}{4}\frac{1-\alpha_g}{d_p}\rho_g\left|\boldsymbol{u}_g - \boldsymbol{u}_p\right|C_{D0}\alpha_g^{-1.65} \tag{4.24}$$

Conversely, for $\alpha_g \leq 0.8$,

$$\beta_{sg} = 150\frac{\left(1-\alpha_g\right)^2\mu_g}{\alpha_g d_p^2} + 1.75\frac{\left(1-\alpha_g\right)\rho_g\left|\boldsymbol{u}_g - \boldsymbol{u}_p\right|}{d_p} \tag{4.25}$$

4.3.4 Syamlal and O'Brien Drag Model

Syamlal and O'Brien (1988) built the drag model based on the correction to the drag coefficient of the single particle which is expressed as

$$\beta_{sg} = \frac{3}{4}\frac{\rho_g}{d_p}\alpha_g\left(1-\alpha_g\right)g\left|\boldsymbol{u}_g - \boldsymbol{u}_p\right|C_{D0}\frac{1}{R_t} \tag{4.26}$$

where R_t is the ratio of the settling velocity of a group of particles and the terminal velocity of a single particle:

$$\begin{aligned}2R_t = {}&C_1\left(\alpha_g\right) - 0.06\,\mathrm{Re}\\&+ \sqrt{\left(0.06\,\mathrm{Re}\right)^2 + 0.12\,Re\left(2C_2\left(\alpha_g\right) - c_1\left(\alpha_g\right) + C_1^2\left(\alpha_g\right)\right)}\end{aligned} \tag{4.27}$$

$$C_1\left(\alpha_g\right) = \alpha_g^{4.14} \tag{4.28}$$

$$C_2\left(\alpha_g\right) = \begin{cases} 0.8\alpha_g^{1.28} & \text{if } \alpha_g < 0.85 \\ \alpha_g^{2.65} & \text{if } \alpha_g \geq 0.85 \end{cases} \tag{4.29}$$

4.3.5   Sub-grid Drag Model

Various drag forces (mentioned in the first paragraph of Sect. 4.3) acting on a group of particles are mainly derived from the force balance relation on the fixed beds or homogeneous liquid-solid fluidized beds, which assumes that the particles are approximately uniformly distributed. In actual gas-solid two-phase flow, the distribution of particle is not uniform in the reactors due to the frequent gas-solid and solid-solid interactions. Although the above drag models consider the effects of the particle concentration on the drag coefficient, however, neglecting the effect of inhomogeneity of particle distribution on the drag force may overestimate the gas-solid drag coefficient, which may make it difficult to accurately capture the multi-scale characteristics of multiphase flow under certain conditions. Thus, the drag forces need to be corrected at the sub-grid level to reasonably reflect the influence of the nonuniformity of particles in the grid.

The effects of meso-scale structures on the drag force can be considered by simply multiplying a nonuniform structure factor H_d with the uniform or nearly uniform particle drag coefficient. Several methods have been proposed in the literature to quantify H_d, among which the filtering grid method and the energy-minimum multi-scale (EMMS) method are two typical approaches. In the filtered grid model, the flow structure is first analyzed using the fine mesh simulation with the mesh size about ten times of the particle diameter or smaller. Then, the relationship between the drag coefficient and the structure-related variables is established, which should be applicable for the simulation on coarse grids. The EMMS model was first proposed by Li and Kwauk [7]. The fluidized system is decomposed into three subsystems which include a particle-rich dense phase, a gas-rich dilute phase, and a dilute and dense interaction phase. The "dense phase" is in the form of clusters or aggregates, the "dilute phase" is in the form of dispersed particles, and the "dilute and dense interaction phase" is a virtual subsystem of the interaction between the dilute and dense phases. For each subsystem, the uniform drag force model is used to describe the gas-solid momentum transfer. Thus, the decomposed system can be described by eight variables, namely, three variables related to the dilute phase (i.e., the gas volume fraction ε_f, superficial gas velocity U_f, and superficial solid velocity U_{pf}), three variables related to the dense phase (the gas volume fraction ε_c, superficial gas velocity U_c, and superficial solid velocity U_{pc}), and two variables describing the interphase (i.e., the cluster size d_{cl} and solid volume fraction f). However, there are only six corresponding EMMS model equations; therefore, the equations do not have a unique solution. To constrain the equations to obtain a unique solution, a stability criterion has been proposed. When the suspended transport energy per unit mass of the particles N_{st} reaches the minimum in the system, it is considered that the gas-solid two-phase flow has reached a stable state. The dilute phase with high gas flow and the dense phase with strong particle interactions tend to be uniform. When the gas phase interacts with the particles, the mesoscale structure has the dominant influence on the flow states causing the nonuniform flow. The details of the model are given in Li and Kwauk [7].

4.4 Turbulence Models

Turbulence models are required to model the turbulent stresses or "Reynolds stresses" in the Reynolds-averaged Navier-Stokes (RANS) equations. Generally, the Boussinesq approximation is employed to determine the turbulent stresses as a function of strain tensor via an eddy viscosity. The shear viscosity of the gas phase μ_g is the sum of the dynamic viscosity μ_{gl} and eddy viscosity μ_{gt}, which is determined by a turbulence model. Various commonly used turbulence models in industrial applications are described here for determining the eddy viscosity μ_{gt}. The transport equations of two-equation shear stress transport (SST) k-ω model and standard k-ε model as well as the one-equation Wray-Agarwal (WA) model and Spalart-Allmaras (SA) model are described below.

4.4.1 Standard k-ε Turbulence Model

As a semiempirical model, the standard κ-ε turbulence model [8], is described by transport equations for the turbulence kinetic energy κ and its dissipation rate ε:

$$\frac{\partial\left(\alpha_g\rho_g k\right)}{\partial t} + \nabla \cdot \left(\alpha_g\rho_g v_g k\right) = \nabla \cdot \left(\alpha_g \frac{\mu_{gt}}{\sigma_k} \nabla k\right) + \alpha_g\left(G_k - \rho_g\varepsilon\right) + \alpha_g\rho_g\Pi_k$$

(4.30)

$$\frac{\partial\left(\alpha_g\rho_g\varepsilon\right)}{\partial t} + \nabla \cdot \left(\alpha_g\rho_g v_g\varepsilon\right) = \nabla \cdot \left(\alpha_g \frac{\mu_{gt}}{\sigma_\varepsilon} \nabla\varepsilon\right) + \alpha_g\left(C_{1\varepsilon}\frac{\varepsilon}{k}G_\kappa - C_{2\varepsilon}\rho_g\frac{\varepsilon^2}{k}\right)$$
$$+ \alpha_g\rho_g\Pi_\varepsilon$$

(4.31)

In Eqs. (4.30) and (4.31), G_κ denotes the generation of k. Π_k and Π_ε represent the interphase turbulent exchange. $C_{1\varepsilon}$ and $C_{2\varepsilon}$ are constants with the values of 1.44 and

1.92, respectively. σ_k and σ_ε are turbulent Prandtl numbers for k and ε, respectively, with $\sigma_k = 1.0$, $\sigma_\varepsilon = 1.3$.

The eddy viscosity is computed as a function of k and ε:

$$\mu_{gt} = \rho_g C_\mu \frac{k^2}{\varepsilon} \tag{4.32}$$

In Eq. (4.32), C_μ is a constant with value 0.09.

4.4.2 SST k-ω Turbulence Model

As a two-equation model, the SST k-ω turbulence model [8] was developed by Menter on the basis of the transport equations for turbulent kinetic energy k and the specific dissipation rate ω. The transport equations are given by

$$\frac{\partial\left(\alpha_g\rho_g k\right)}{\partial t} + \nabla \cdot \left(\alpha_g\rho_g v_g k\right) = \nabla \cdot \left[\alpha_g\left(\mu_{gl} + \frac{\mu_{gt}}{\sigma_k}\right)\nabla k\right] + \alpha_g G_k - \alpha_g\rho_g\beta^*\omega k \tag{4.33}$$

$$\frac{\partial\left(\alpha_g\rho_g\omega\right)}{\partial t} + \nabla \cdot \left(\alpha_g\rho_g v_g\omega\right) = \nabla \cdot \left[\alpha_g\left(\mu_{gl} + \frac{\mu_{gt}}{\sigma_\omega}\right)\nabla\omega\right] + \alpha_g G_\omega - \alpha_g\rho_g\beta\omega^2$$
$$+ 2(1 - F_1)\frac{\alpha_g\rho_g}{\sigma_{\omega,2}\omega}\nabla k \cdot \nabla\omega \tag{4.34}$$

where G_k and G_ω denote the production of k and ω, respectively. The turbulent Prandtl numbers for k and ω are σ_k and σ_ω, respectively, which are expressed as follows:

$$\sigma_k = \frac{1}{F_1/\sigma_{k,1} + (1 - F_1)/\sigma_{k,2}} \tag{4.35}$$

$$\sigma_\omega = \frac{1}{F_1/\sigma_{\omega,1} + (1 - F_1)/\sigma_{\omega,2}} \tag{4.36}$$

$\sigma_{k,\,1}$, $\sigma_{k,\,2}$, $\sigma_{\omega,\,1}$, and $\sigma_{\omega,\,2}$ are constants with the values of 1.176, 1.0, 2.0, and 1.168, respectively.

The blending function F_1 is expressed as

$$F_1 = \tanh\left\{\left\{\min\left[\max\left(\frac{\sqrt{k}}{0.09\omega d}, \frac{500\mu}{\rho d^2\omega}\right), \frac{4\rho_g k}{\sigma_{\omega,2}D_\omega^+ d^2}\right]\right\}^4\right\} \tag{4.37}$$

$$D_\omega^+ = \max\left(2\rho_g \frac{1}{\sigma_{\omega,2}} \frac{1}{\omega} \nabla k \cdot \nabla \omega, 10^{-10}\right) \tag{4.38}$$

where D_ω^+ represents the positive portion of the cross-diffusion term.

The eddy viscosity μ_{gt} for SST k-ω model is given by

$$\mu_{gt} = \frac{\rho_g k}{\omega} \frac{1}{\max\left[\frac{1}{\alpha^*}, \frac{SF_2}{a_1 \omega}\right]} \tag{4.39}$$

$$F_2 = \tanh\left\{\left[\max\left(2\frac{\sqrt{k}}{0.09\omega d}, \frac{500\mu}{\rho_g d^2 \omega}\right)\right]^2\right\} \tag{4.40}$$

where S is the magnitude of strain rate and a_1 is a constant $= 0.31$.

4.4.3 Wray-Agarwal (WA) Turbulence Model

The WA turbulence model [9] is a one-equation model which is derived from the k-ω model and the transport equation for R ($R = k/\omega$). WA 2017m version of the model is given below [9]:

$$\frac{\partial R}{\partial t} + \nabla \cdot (v_g R) = \nabla \cdot \left[(\sigma_R R + v_g)\nabla R\right] + C_1 R S + f_1 C_{2k\omega}\frac{R}{S}\nabla R \cdot \nabla S$$
$$- (1 - f_1)C_{2k\varepsilon} \min\left[R^2\left(\frac{\nabla S \cdot \nabla S}{S^2}\right), C_m \nabla R.\nabla R\right] \tag{4.41}$$

where v_g is the kinematic viscosity; C_1, $C_{2k\omega}$, $C_{2k\varepsilon}$, and σ_R are constants; S represents the mean strain given below:

$$S = \sqrt{2S_{ij}S_{ij}}, \qquad S_{ij} = \frac{1}{2}\left(\frac{\partial v_{gi}}{\partial x_j} + \frac{\partial v_{gj}}{\partial x_i}\right) \tag{4.42}$$

The switching function f_1 is expressed as

$$f_1 = \min\left(\tanh\left(\arg_1^4\right), 0.9\right) \tag{4.43}$$

$$\arg_1 = \frac{1 + \frac{d\sqrt{RS}}{v}}{1 + \left[\frac{\max\left(d\sqrt{RS}, 1.5R\right)}{20v}\right]^2} \tag{4.44}$$

where d is the distance from the nearest wall.

The constants in the transport equation are given as

$$C_{1k\omega} = 0.0829 \quad C_{1k\varepsilon} = 0.1127$$

$$C_1 = f_1(C_{1k\omega} - C_{1k\varepsilon}) + C_{1k\varepsilon}$$

$$\sigma_{k\omega} = 0.72 \quad \sigma_{k\varepsilon} = 1.0$$

$$\sigma_R = f_1(\sigma_{k\omega} - \sigma_{k\varepsilon}) + \sigma_{k\varepsilon}$$

$$k = 0.41$$

$$C_{2k\omega} = \frac{C_{1k\omega}}{k^2} + \sigma_{k\omega} \quad C_{2k\varepsilon} = \frac{C_{1k\varepsilon}}{k^2} + \sigma_{k\varepsilon} \tag{4.45}$$

The turbulence eddy viscosity is expressed as

$$\mu_{gt} = \rho_g f_\mu R \tag{4.46}$$

$$f_\mu = \frac{\chi^3}{\chi^3 + C_\omega^3} \quad \chi = \frac{R}{\nu_g} \tag{4.47}$$

where f_μ is the damping function which describes the wall blocking effect and C_m is a constant with the value of 8.54.

4.4.4   *Spalart-Allmaras (SA) Turbulence Model*

The Spalart-Allmaras (SA) turbulence model is an industry standard one-equation linear eddy viscosity turbulence model originally developed for the prediction of aerodynamic flows [10]. It is a low-Reynolds number model and requires the boundary layer near the wall for accuracy. The SA model is given by the following equation:

$$\begin{aligned}
\frac{\partial \tilde{v}}{\partial t} + u_j \frac{\partial \tilde{v}}{\partial x_j} &= c_{b1}(1 - f_{t2})\tilde{S}\tilde{v} - \left[c_{w1}f_w - \frac{c_{b1}}{\kappa^2}f_{t2}\right]\left(\frac{\tilde{v}}{d}\right)^2 + \frac{1}{\sigma} \\
&\quad \times \left[\frac{\partial}{\partial x_j}\left\{(v + \tilde{v})\frac{\partial \tilde{v}}{\partial x_j}\right\} + c_{b2}\frac{\partial \tilde{v}}{\partial x_i}\frac{\partial \tilde{v}}{\partial x_i}\right]
\end{aligned} \tag{4.48}$$

The turbulent eddy viscosity is given by the equation

$$\mu_t = \rho \tilde{v} f_{v1} \tag{4.49}$$

Near wall blocking is accounted for by the damping function f_{v1}:

$$f_{v1} - \frac{\chi^3}{\chi^3 + c^3{}_{v1}} \tag{4.50}$$

$$\chi = \frac{\tilde{v}}{v} \tag{4.51}$$

The remaining functions are defined as follows:

$$\tilde{S} = \Omega + \frac{\tilde{v}}{\kappa^2 d^2} f_{v2}, \quad f_{v2} = 1 - \frac{\chi}{1 + \chi f_{v1}} \tag{4.52}$$

where $\Omega = \sqrt{2 W_{ij} W_{ij}}$ is the magnitude of the vorticity and d is the distance of the flow field point to the nearest wall. W_{ij} is defined by the following equation:

$$W_{ij} = \frac{1}{2} \left(\frac{\partial u_i}{\partial x_j} - \frac{\partial u_j}{\partial x_i} \right) \tag{4.53}$$

$$f_\omega = g \left[\frac{1 + c_{\omega3}^6}{g^6 + c_{\omega3}^6} \right]^{\frac{1}{6}}, \quad g = r + c_{\omega2}\left(r^6 - r\right), \quad r = \min\left[\frac{\tilde{v}}{\tilde{S}\kappa^2 d^2}, 10 \right] \tag{4.54}$$

$$f_{t2} = c_{t3} \exp\left(- c_{t4}\chi^2\right) \tag{4.55}$$

The model constants are

$$c_{b1} = 0.1355, \quad \sigma = \frac{2}{3} \quad c_{b2} = 0.622, \quad \kappa = 0.41, \quad c_{\omega2} = 0.3,$$

$$c_{\omega3} = 2, \quad c_{v1} = 7.1, \quad c_{t3} = 1.2, \quad c_{t4} = 0.5, \quad c_{w1} = \frac{c_{b1}}{\kappa^2} + \frac{1 + c_{b2}}{\sigma} \tag{4.56}$$

4.5 Chemical Kinetics Models

The macroscopic gas-solid reaction includes several microcosmic intermediate steps [11]. The gas reactants first diffuse from the bulk of the gas phase to the surface of the reacting solid particle in the film diffusion process. Then, the gaseous reactants flow through the pores of the particle and/or the product layer formed in the surface. The reactants will further be adsorbed and gas-solid reaction occurs. The gas products formed in the reaction will flow via the opposite directions, which is desorbed from the solid particle and diffuse to the bulk of the surface. However, the gas-solid reaction may also be affected by the heat transfer and solid structure changes such as sintering and agglomeration.

A reliable theoretical model is beneficial for in-depth understanding of the reaction behaviors in CLC process and even provides valuable guidance for the design of oxygen carrier. The reactions of oxygen carrier in CLC can be considered as non-catalytic gas-solid reactions which are given as

$$a\mathrm{A}(g) + \mathrm{B}(s) \rightarrow b\mathrm{P}(s) + c\mathrm{CO}_2(g) + d\mathrm{H}_2\mathrm{O}(g) \tag{4.57}$$

Currently, the frequently used gas-solid reaction models are the changing grain size model (CGSM), the shrinking core model (SCM), and the nucleation and nuclei growth model (NNGM).

4.5.1   Shrinking Core Model (SCM)

The shrinking core model (SCM) can be considered for the gas-solid flow when the resistance to the gas diffusion in the unreacted particle is high. As illustrated in Fig. 4.2, in the SCM, the solid product is formed in the outer of a particle which surrounds the unreacted core. A clear shrinking interface is defined between the nonporous unreacted core and porous product layer. The gas-solid reaction initially occurs in the particle outer surface. The interface is then formed where the reaction happens. It gradually moves inward, and the unreacted core diminishes in size with the proceeding of the reaction. The mass transfer equation considering the diffusion through the product layer is shown below:

$$\frac{d}{dr}\left(-D_{e,A}r^2\frac{dC_A}{dr}\right) = 0, \quad C_A(r)|_{r=r_0} = C_{A0}, \quad C_A(r)|_{r=r_2} = 0 \tag{4.58}$$

where r is the radius direction, C_A is the concentration of gas species A, $D_{e,A}$ is the gas diffusion coefficient, C_{A0} is concentration of gas species A in the bulk, r_0 is initial radius of the particle, and r_2 is radius of the unreacted core.

The radius of the unreacted core r_2 is expressed as

$$\frac{dr_2}{dt} = -\frac{aC_{A0}/\rho_B}{\frac{r_2^2}{r_0^2 k_g} + \frac{(r_0 - r_2)r_2}{r_0 D_c} + \frac{1}{k_s}} \tag{4.59}$$

where the three terms in the denominator are the resistance of gas-film diffusion, product layer diffusion, and the chemical reaction, respectively. The reacting gas should overcome the above three resistances before the gas-solid reaction happens.

Ishida et al. [12] applied the SCM to predict the kinetic parameters in the reduction and oxidation of Ni-based oxygen carrier. The results show that the reduction process was controlled by chemical reaction, while the oxidation process was dependent on both of the product layer diffusion and chemical reaction. The SCM model is able to calculate the reaction rate of small particles with relatively high accuracy [13, 14] but shows uncertainty for the particles with high porosity and size of about 100 mm.

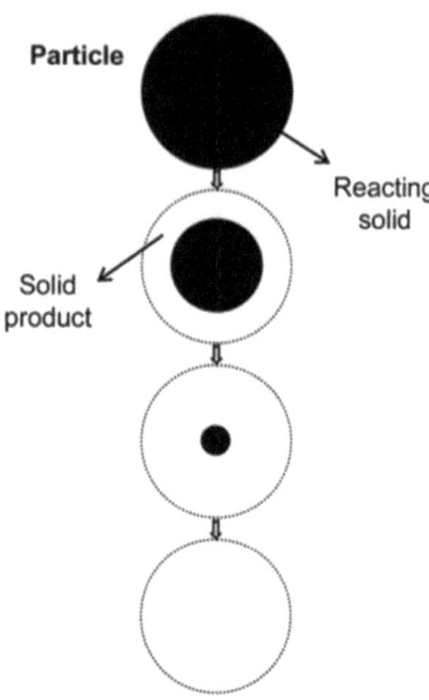

Fig. 4.2 Schematic diagram of shrinking core model [15]

4.5.2 *Changing Grain Size Model (CGSM)*

The changing grain size model (CGSM) assumes that the active materials are in the form of fine and nonporous grains distributed in the porous particle. Each grain grows or shrinks following the SCM when the reaction happens, as displayed in Fig. 4.3. The update of the overall heterogeneous reaction rate and particle porosity occurs with the change of the grain size.

Considering the gas-solid reaction, the unsteady mass transfer equation in CGSM is expressed as

$$\frac{\partial C_A}{\partial t} = \frac{1}{r^2} \frac{\partial}{\partial R} \left(D_{e,A} R^2 \frac{\partial C_A}{\partial R} - (-r_A) \right) \tag{4.60}$$

where $-r_A$ is the reaction rate of gas species A and C_A is the gas concentration at the radial position R.

The boundary conditions are

$$C_A(R,t)|_{t=0} = C_{A,b} \tag{4.61}$$

$$\left. \frac{dC_A}{dR} \right|_{R=0} = 0 \tag{4.62}$$

$$-D_{e,A} \left. \frac{dC_A}{dR} \right|_{R=R0} = k_g (C_{A,s} - C_{A,b}) \tag{4.63}$$

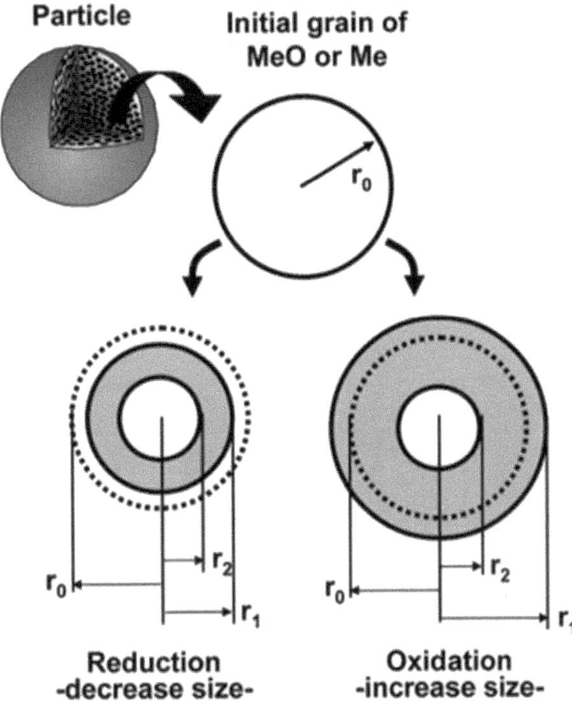

Fig. 4.3 Changing grain size model [15]

where $C_{A,b}$ and $C_{A,s}$ are the gas concentration in the bulk phase and external surface of the solid particle, respectively. k_g is the external mas transfer coefficient.

The reaction rate per unit of particle volume is

$$\frac{1}{a}(-r_A) = -r_R = kS_{0,R}\left(\frac{r_2}{r_0}\right)^2 C_A \tag{4.64}$$

where r_0 and r_2 are the radius of the grain and the unreacted core, respectively.

4.5.3   Nucleation and Nuclei Growth Models (NNGM)

The nucleation and nuclei growth model (NNGM) assumes the reaction proceeds by the generation of nucleation (nuclei formation) and subsequent nuclei growth. As displayed in Fig. 4.4, the solid phase is first activated to form nuclei, which is called induction period. The length of the induction period depends on the reaction temperature and gas-solid type. The reaction happens when the nuclei is formed. With the continuous nucleation and growth of the formed nuclei, the reaction rate gradually increases. The nuclei grow via the overlapping of the nuclei and ingestion of nucleation sites. The reaction happens uniformly over the surface of the solid particle and gradually advances into the core of the grain. The overall reaction rate is

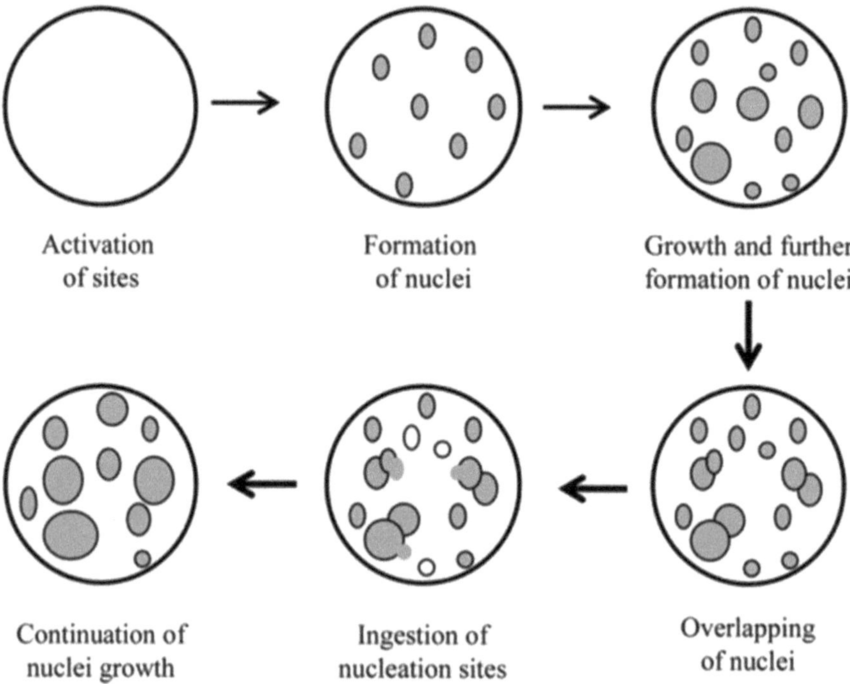

Fig. 4.4 Nucleation and nuclei growth model [15]

dependent on the nucleation rate, nuclei growth, and germ nuclei, which is further determined by the temperature and the gas components. The rate-determining step is the nucleation rate and nuclei growth rate. In the NNGM, the conversion can be described by the Avrami-Erofeev Model (AEM) [16]:

$$\frac{dX_s}{dt} = k'_s(T)C_g^n f(X_s) \tag{4.65}$$

$$f(X_s) = \nu(1 - X_s)[-\ln(1 - X_s)]^{(\nu-1)/\nu} \tag{4.66}$$

where $f(X_s)$ is a function of the solid conversion; n is the Avrami exponent indicative of the reaction mechanism and crystal growth dimension. The random nucleation model (RNM) is given by a value of $\nu = 1$ in which the induction period is not present. $\nu = 2$ and 3 represent the two-dimensional or three-dimensional nuclei growth, respectively.

The AEM method is used to describe the reduction and oxidation processes of Ni-based oxygen carrier [16]. The kinetic parameters were obtained from the temperature-programmed reduction or oxidation (TPR/TPO) profile analysis. The reduction reaction happens under the temperature of 300–600 °C with a heating rate of 10 °C/min, while the oxidation process occurs under the lower temperature of 200–500 °C.

The nucleation effects described in the NNGM method are very obvious in the metallic oxides reduction [17], and the NNGM has been well validated in the Ni-based catalyst reduction at low temperature. The induction period before the nucleation is clearly observed with the solid conversion lower than 0.1–0.2 and temperature lower than 300 °C [11]. With the increase of temperature, the induction process becomes less evident, and it becomes imperceptible at the temperature of 340 °C due to the acceleration of nucleation process. Sedor et al. [18] found that the reaction starts immediately, with no induction period, at temperatures above 600 °C. The nucleation process is fast due to the high temperature of over 800 °C in the CLC process, and the solid conversion is not greatly affected by the temperature distribution. The model describing the interfacial chemical reaction can be used when the nuclei is formed all over the entire solid surface, and SCM shows good prediction.

4.5.4   Kinetic Data

In the modeling process, it is essential to determine the kinetics including the activation energy and reaction order for different reaction schemes. The effects of temperature and gas concentration should be adequately measured. Table 4.1 gives the kinetic data for the oxidation of Cu-based, Ni-based, and Fe-based oxygen carriers with oxygen and reduction of the oxygen carriers with the reducing gases such as CH_4, CO, and H_2. Most of the kinetics are obtained from the experiments carried out in thermogravimetric analyzers (TGA). The temperature-programmed reduction and/or oxidation (TPR, TPO) is used for the determination of kinetic data at a low temperature range, and the external mass transfer is not considered. Fluidized bed and fixed bed reactors can also be used for obtaining the kinetics, but it is essential to reduce the limitation of mass transfer or use a reaction model which has considered the mass transfer process.

In Table 4.1, it can be seen the reaction orders are in the range of 0.8–1.0 in general, and the activation energy follows the overall tendency of $CH_4 > H_2 > CO \approx O_2$. The obtained kinetics vary for different types oxygen carrier. The synthetic oxygen carriers with metal oxides supported on an inert material are widely used in the experiments, and the interaction between the metal oxide and support affects the activation energy. Only a few experiments investigate the kinetics of natural minerals, such as ilmenite (Fe_2TiO_5) and anhydrite ($CaSO_4$). In the CLC process, when the natural gas, syngas, or coal is used, the CO and H_2 are the intermediate gases. Thus, it is necessary to know the kinetic data for the reaction between one specific type of oxygen carrier and different reducing gases, which can also be found in Table 4.1. In addition, the kinetics for the oxidation process can be used for the design and optimization of the air reactor.

Table 4.1 Summary of kinetic data determined for oxygen carrier [15]

Oxygen carrier	Experimental condition	Kinetic model	Reference
60 wt% NiO on YSZ	TGA $T = 800–1000$ °C 100 vol% H_2 21 vol% O_2	SCM $n = 1$, $E_r = 82$ kJ/mol $n = 1.0$, $E_r = 17$ kJ/mol	[12]
58 wt% NiO on bentonite	TGA $T = 600–750$ °C 5 vol% CH_4	SCM $E_r = 37$ kJ/mol	[13]
60 wt% NiO on Al_2O_3	TGA $T = 600–950$ °C 5–70 vol% CH_4 5–70 vol% H_2 5–70 vol% CO 5–70 vol% O_2	SCM $n = 0.8$, $E_r = 78$ kJ/mol $n = 0.5$, $E_r = 26$ kJ/mol $n = 0.8$, $E_r = 25$ kJ/mol $n = 0.2$, $E_r = 7$ kJ/mol	[19, 20]
60 wt% CuO on SiO_2	TGA $T = 700–850$ °C 100 vol% CH_4	SCM $n = 1$, $E_r = 41$ kJ/mol	[21]
10 wt% CuO on Al_2O_3	TGA $T = 600–800$ °C 5–70 vol% CH_4 5–70 vol% H_2 5–70 vol% CO 5–21 vol% O_2	SCM $n = 0.4$, $E_r = 60$ kJ/mol $n = 0.6$, $E_r = 33$ kJ/mol $n = 0.8$, $E_r = 14$ kJ/mol $n = 1.0$, $E_r = 15$ kJ/mol	[22]
60 wt% Fe_2O_3 on Al_2O_3	TGA $T = 600–950$ °C 5–70 vol% CH_4 5–70 vol% H_2 5–70 vol% CO 5–21 vol% O_2	SCM $n = 1.3$, $E_r = 49$ kJ/mol $n = 0.5$, $E_r = 24$ kJ/mol $n = 1.0$, $E_r = 20$ kJ/mol $n = 1.0$, $E_r = 14$ kJ/mol	[19, 20]
Calcined ilmenite (Fe_2TiO_5)	TGA $T = 800–850$ °C 5–50 vol% CH_4 5–50 vol% H_2 5–50 vol% CO 5–50 vol% O_2	SCM $n = 1.0$, $E_r = 165$ kJ/mol $n = 1.0$, $E_r = 109$ kJ/mol $n = 1.0$, $E_r = 113$ kJ/mol $n = 1.0$, $E_r = 12$ kJ/mol	[23]
$CaSO_4$	TPR $T = 850–1200$ °C 20 vol% CO	AEM with $\gamma = 2$ $E_r = 280$ kJ/mol	[24]
$CaSO_4$	Fixed bed $T = 880–950$ °C 20–70 vol% CO	SCM $E_r = 145$ kJ/mol	[25]

4.6   Softwares for Multiphase Flow Simulations

ANSYS Fluent is a commercial computational fluid dynamics tool which includes well-validated physical modeling capabilities for a wide range of incompressible and compressible, laminar and turbulent fluid, and steady and transient flow patterns. Accurate simulation can be conducted for different types of multiphase flows such as gas-solid, gas-fluid, fluid-fluid, and particle flows. Accurate reaction prediction requires the precise mix of a broad range of mathematical models including chemical reactions, heat transfer, momentum transfer, and so on. ANSYS Fluent also provides access to add new functions to the existing capability via the development and implementation of the user-defined function (UDF), but the process is tedious.

Barracuda Virtual Reactor is a commercial computational particle fluid dynamics (CPFD) software. Since the modeling of particles as fluids is inherently limited in CFD software, Barracuda is developed to model the behaviors of particles in industrial equipment, which makes the industrial design more efficient and environmentally friendly. The multiphase particle-in-cell (MP-PIC) approach is used in the CPFD software. The MP-PIC method groups the particles with the same velocities and properties into computational particles named parcels. The interactions between parcels instead of the particles are calculated, which greatly reduces the computational cost. Barracuda is widely used by major automotive and transportation manufacturers, as well as by major chemical, petrochemical, and power generation companies. They continue to expand the application to meet more critical energy challenges.

Multiphase Flow with Interphase eXchanges (MFiX) is an open-source software for multiphase flow simulation which is developed by the National Energy Technology Laboratory (NETL). It has over two decades of development history and has attracted more than 7000 registrations until 2021. Compared with commercial software, MFiX provides the source code in Fortran programming language for users to modify or add new function to solve particular problems. MFiX has been applied to a diverse range simulation of multiphase flow and reactions in bubbling fluidized bed, spouted fluidized bed, cyclone, etc. Different numerical schemes are available in MFiX including the Eulerian-Eulerian method, discrete element method (DEM), and particle-in-cell (PIC) method. MFiX has become the standard for comparing, implementing, and evaluating multiphase flow constitutive models.

OpenFOAM is another open-source software for multiphase flow and reaction simulation. It was released in 2004 and developed on Linux operating system. C++ programming language is used with the advantages of modularity and extensibility. It is easy for users to develop, test, or validate new capabilities. New OpenFOAM version is continuously released by updating or adding the numerical algorithms, turbulence models, grid types, and so on.

References

1. J. Jung, I.K. Gamwo, Multiphase CFD-based models for chemical looping combustion process: fuel reactor modeling. Powder Technol. **183**(3), 401–409 (2008)
2. C.K.K. Lun, S.B. Savage, D.J. Jeffrey, N. Chepurniy, Kinetic theories for granular flow: inelastic particles in couette flow and slightly inelastic particles in a general flowfield. J. Fluid Mech. **140**, 223–256 (1984)
3. D. Gidaspow, D. Gidaspow, Multiphase flow and fluidization: continuum and kinetic theory descriptions. AICHE J. **42**(4), 1197–1198 (1994)
4. D.J. Gunn, Transfer of heat or mass to particles in fixed and fluidised beds. Int. J. Heat Mass Transf. **21**(4), 467–476 (1978)
5. J.M. Link, *Development and Validation of a Discrete Particle Model of a Spout-Fluid Bed Granulator* (University of Twente, Enschede, 1975)
6. Guide, A. F. U. *ANSYS Fluent Theory Guide* (ANSYS Inc., Nov 2013)

7. L. Jing, hai, K. M., *Particle-Fluid Two-Phase Flow the Energy-Minimization Multi-Scale Method* (Beijing Metallurgical Industry Press, 1994)
8. ANSYS Inc (Release-15.0). *ANSYS Fluent Theory Guide* (ANSYS Inc., USA, 2015), *15317* (January), 1–759
9. H. Xu, T.J. Wray, R.K. Agarwal Application of a new DES model based on Wray-Agarwal turbulence model for simulation of wall-bounded flows with separation. AIAA Paper 2017-3966, *AIAA 47th Fluid Dynamics Conferences, Denver, CO,* 5–9 June 2017
10. P. Spalart, S. Allmaras, A one-equation turbulence model for aerodynamic flows. AIAA Paper 1992-439, *30th AIAA Aerospace Sciences Meeting and Exhibit, Reno, NV,* 6–9 January 1992
11. J. Szekely, C.I. Lin, H.Y. Sohn, A structural model for gas—solid reactions with a moving boundary—V an experimental study of the reduction of porous nickel-oxide pellets with hydrogen. Chem. Eng. Sci. **28**(11), 1975–1989 (1973)
12. M. Ishida, H. Jin, T. Okamoto, A fundamental study of a new kind of medium material for chemical-looping combustion. Energy Fuel **10**(4), 958–963 (1996)
13. H.-J. Ryu, D.-H. Bae, K.-H. Han, S.-Y. Lee, G.-T. Jin, J.-H. Choi, Oxidation and reduction characteristics of oxygen carrier particles and reaction kinetics by unreacted core model. Korean J. Chem. Eng. **18**(6), 831–837 (2001)
14. S.R. Son, S.D. Kim, Chemical-looping combustion with NiO and Fe2O3 in a thermobalance and circulating fluidized bed reactor with double loops. Ind. Eng. Chem. Res. **45**(8), 2689–2696 (2006)
15. J. Adanez, A. Abad, F. Garcia-Labiano, P. Gayan, L.F. De Diego, Progress in chemical-looping combustion and reforming technologies. Prog. Energy Combust. Sci. **38**(2), 215–282 (2012)
16. M.M. Hossain, H.I. de Lasa, Reactivity and stability of Co-Ni/Al2O3 oxygen carrier in multicycle CLC. AICHE J. **53**(7), 1817–1829 (2007)
17. J. Szekely, J.W. Evans, H.Y. Sohn, *Gas-Solid Reactions* (Academic Press Inc, New York, 1976)
18. K.E. Sedor, M.M. Hossain, H.I. de Lasa, Reduction kinetics of a fluidizable nickel–alumina oxygen carrier for chemical-looping combustion. Can. J. Chem. Eng. **86**(3), 323–334 (2008)
19. A. Abad, J. Adánez, F. García-Labiano, L.F. de Diego, P. Gayán, J. Celaya, Mapping of the range of operational conditions for Cu-, Fe-, and Ni-based oxygen carriers in chemical-looping combustion. Chem. Eng. Sci. **62**(1–2), 533–549 (2007)
20. A. Abad, F. Garcia-Labiano, L.F. de Diego, P. Gayán, J. Adánez, Reduction kinetics of Cu-, Ni-, and Fe-based oxygen carriers using syngas (CO+ H2) for chemical-looping combustion. Energy Fuel **21**(4), 1843–1853 (2007)
21. J. Adáñez, F. Garcí́a-Labiano, L.F. de Diego, A. Plata, J. Celaya, P. Gayá́n, A. Abad, Optimizing the fuel reactor for chemical looping combustion. In *International Conference on Fluidized Bed Combustion* Vol. 36800, pp 173–182 (2003)
22. F. García-Labiano, L.F. de Diego, J. Adánez, A. Abad, P. Gayán, Reduction and oxidation kinetics of a copper-based oxygen carrier prepared by impregnation for chemical-looping combustion. Ind. Eng. Chem. Res. **43**(26), 8168–8177 (2004)
23. A. Abad, J. Adanez, A. Cuadrat, F. Garcia-Labiano, P. Gayan, F. Luis, Kinetics of redox reactions of ilmenite for chemical-looping combustion. Chem. Eng. Sci. **66**(4), 689–702 (2011)
24. H. Tian, Q. Guo, Investigation into the behavior of reductive decomposition of calcium sulfate by carbon monoxide in chemical-looping combustion. Ind. Eng. Chem. Res. **48**(12), 5624–5632 (2009)
25. R. Xiao, Q.L. Song, W.G. Zheng, Z.Y. Deng, L.H. Shen, Zhang, M.Y., Reduction kinetics of a CasO4 based oxygen carrier for chemical-looping combustion. In *Proceedings of the 20th International Conference on Fluidized Bed Combustion*, Springer, pp 519–526 (2009)

Chapter 5
Eulerian Simulation of a CLC Reactor

5.1 Description of the Experimental Setup

The experiment employs the two-compartment fluidized bed design proposed by Chong et al. [1] and further investigated by Fang et al. [2]; the experimental setup is shown in Fig. 5.1. Dimensions and additional details can be found in Abad et al. [3] The oxygen carrier particles in the air reactor are oxidized in the presence of air; the fluidizing velocity is greater than the terminal velocity of the particles and carries the particles upward. The flow then undergoes a sudden expansion in the particle separator at the top of the reactor, which causes the particles to fall back down into the downcomer and enter the fuel reactor.

The experiment employed a Fe-based oxygen carrier consisting of 60% Fe_2O_3 by mass and 40% Al_2O_3. The analysis of several metal oxides and alloys showed that this oxygen carrier provides excellent reactivity for use in CLC, and its hardness and resistance to agglomeration are ideal for fluidized bed operation [4]. The Al_2O_3 is inert and acts as a porous support providing a higher surface area for reaction. The particular batch of oxygen carrier used by Abad et al. [3] was sintered at $1100\,^{\circ}C$ and is designated as F6A1100. The gaseous fuels used in the experiment are natural gas consisting primarily of CH_4 and syngas simulated by a mixture of 50% CO and 50% H_2. The fluidizing velocity in the fuel reactor is below the terminal velocity of the particles; hence a bubbling bed behavior is exhibited. Therefore, the particles do not reach the particle separator in the fuel reactor. The pressure in the fuel reactor is controlled via a water trap connected to the flue stream of the reactor to ensure minimal gas leakage between the fuel reactor and the air reactor through the downcomer and the slot. The flue streams from both reactors are led to a gas analyzer where the concentrations of various gases are measured.

R. K. Agarwal, Y. Shao, *Modeling and Simulation of Fluidized Bed Reactors for Chemical Looping Combustion*, https://doi.org/10.1007/978-3-031-11335-2_5

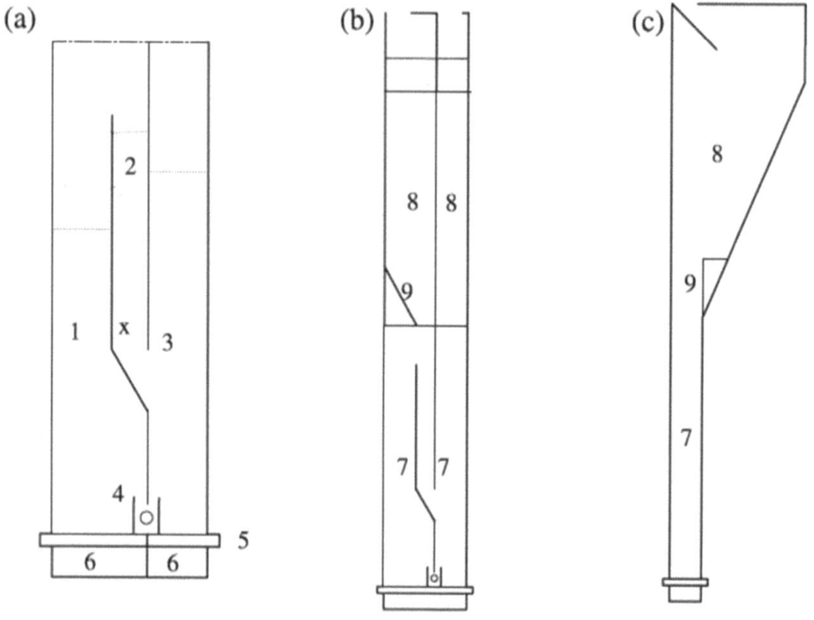

Fig. 5.1 Sketch of experimental reactor: (*1*) air reactor, (*2*) downcomer, (*3*) fuel reactor, (*4*) slot, (*5*) gas distributor plate, (*6*) wind box, (*7*) reactor part, (*8*) particle separator, (*9*) leaning wall. The symbols (*x*) and (*o*) indicate fluidization in the downcomer and slot

5.2   Chemical Reactions and Rates

The metal oxide reduction reactions used in the simulation are

$$3\,Fe_2O_3 + CO \rightarrow 2\,Fe_3O_4 + CO_2 \tag{5.1}$$

$$3\,Fe_2O_3 + H_2 \rightarrow 2\,Fe_3O_4 + H_2O \tag{5.2}$$

Exact reaction rates for the reduction of F6A1100 with CO and H_2 are not available in the literature; the reaction rates are assumed to be the same as the reduction rates for hematite (pure Fe_2O_3) with CO and H_2 obtained from the experimental work of Mattisson et al. [5] and further developed by Mahalatkar et al. [6] for the simulation of chemical reactions in a single reactor setup with solid fuel. Based on these papers, the reaction rates $\dot{m}$ (in kg/s per cell volume or kg/(m^{3}s)) of the fuel gases with iron oxide are given by

$$\dot{m}_{H_2} = \frac{k_{H_2} R_o}{2MW_{O_2}} \rho_{avg} \alpha_s \left(Y_{Fe_2O_3} + Y_{Fe_3O_4} \frac{\nu_{Fe_2O_3} MW_{Fe_2O_3}}{\nu_{Fe_3O_4} MW_{Fe_3O_4}} \right) (1-X)^{2/3} MW_{H_2} \quad (5.3)$$

$$\dot{m}_{CO} = \frac{k_{CO} R_o}{2MW_{O_2}} \rho_{avg} \alpha_s \left(Y_{Fe_2O_3} + Y_{Fe_3O_4} \frac{\nu_{Fe_2O_3} MW_{Fe_2O_3}}{\nu_{Fe_3O_4} MW_{Fe_3O_4}} \right) (1-X)^{2/3} MW_{CO} \quad (5.4)$$

where k is the nominal reaction rate based on the Arrhenius rate, R_o is the oxygen-carrying capacity, MW is the molecular weight (in kg/kmol), Y is the mass fraction, ν is the stoichiometric coefficient, and X is the conversion fraction based on the fully reduced state; in each case, the subscript identifies the species under consideration. More details of the reaction rate derivation can be found in Mahalatkar et al. [6] The reaction rates identified in Eqs. (5.3) and (5.4) are implemented into the numerical simulation through separate user-defined functions.

5.3   Two-Dimensional Simulation of Experiment of Abad et al. [3]

The simple rectilinear geometry of the reactor used in the experiment of Abad et al. [3] can be easily modeled using a 2D setup. The absence of cylindrical elements such as a cyclone separator suggests that the introduction of swirl velocities and other 3D effects will be minimal. The numerical grid used in the CFD simulation is based on the experimental setup shown in Fig. 5.1 with few changes. Since the simulation assumes a 2D domain, the expansion in the particle separator is modified to take place in the plane of the reactor. Since the expansion takes place downstream of the circulating bed, it is expected that the bed behavior remains unaffected. The leaning wall and the wind box are also removed, and the fluidizing gas is introduced directly at the bottom of the air and fuel reactors. The resulting geometry is meshed with a relatively fine grid in the lower regions of the geometry where the solid circulation occurs (particularly in the slot region) and a coarser grid in the upper regions where the expansion takes place. This ensures that an accurate solution for the fluidization behavior is obtained while also limiting the computational cost. The final CFD mesh is shown in Fig. 5.2.

The experimental reactor of Abad et al. [3] was operated for 60 h with both natural gas and syngas without replacing the oxygen carrier particles or adding new material. Given the time taken to complete 20 s of simulation on a Dell workstation with a quad-core Intel Xeon central processing unit (approximately 48 h), the complete reactor simulation of 60 h is beyond the scope of CFD at this time. Instead, the initial batch processing results of Abad et al. [3] are used to validate the CFD simulation reported here. For these batch operations, which last less than 1 min, the initial oxygen carrier mass in the fuel reactor is sufficient for reacting with all the incoming fuel; thus the fuel conversion is not affected by the oxygen carrier

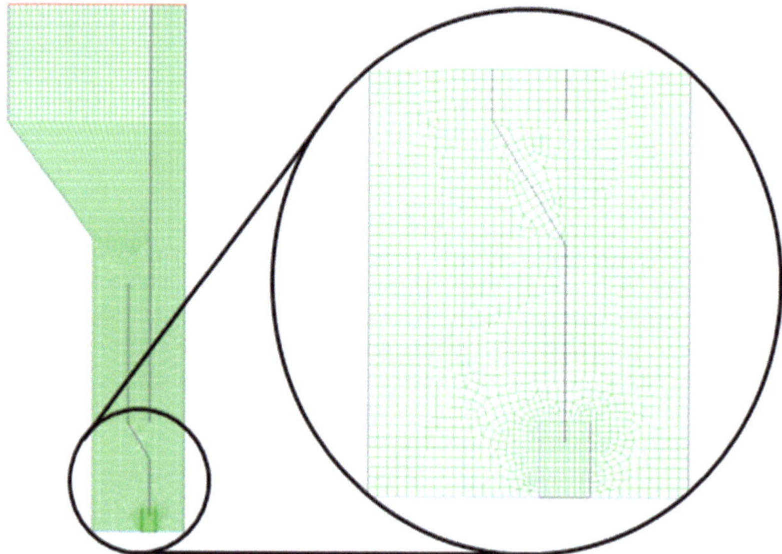

Fig. 5.2 Computational domain and grid for 2-D CFD simulation with detailed view of lower bed

re-oxidation in the air reactor. The CFD simulation is thus considerably simplified by setting the fluidization gas in the air reactor to an inert gas (nitrogen in this case). Of the two gaseous fuels used in the experiment, only syngas has been considered in the current work because the chemical kinetics for the reaction of Fe_2O_3 with the non-methane components of natural gas are not available.

The oxygen carrier used in the experiment has an apparent density of 2150 kg/m^3 and porosity of 0.56 with a diameter of 90–212 μm; the average value of 150 μm is used in the simulation. The initial bed height is 10 cm, corresponding to a bed mass of 180 g. The reactors are set to atmospheric pressure and the gage pressures at the outlets are set at zero. The pressure differential between the reactors, controlled by a water trap in the experiment to minimize leakage, was not implemented in the simulation because the data was not available. It is expected that the pressure differential is a secondary mechanism and the dense solid packing in the downcomer and the slot will be sufficient to keep the leakage to a minimum. Lastly, the temperature for the simulation was set at 1123 K, in line with the reference condition specified from the experiment. The numerical parameters used in the CFD simulation are summarized in Table 5.1. It should be noted that the secondary phase mass fraction has been set to zero at both fuel and air reactor inlets, i.e., no new oxygen carrier is added.

The simulation was run for approximately 20 s. Figure 5.3 shows the instantaneous contour plots of the solid volume fraction during the initial development of solids flow in the dual-fluidized bed reactor system for the first 1 s. Initially, the oxygen carrier particles in the air reactor are lifted by the fluidizing velocity until they reach the particle separator. The reduction in flow velocity due to the expansion

Table 5.1 Key modeling parameters for CFD simulation of Abad et al.'s experiment [3]

Primary phase	Fuel-gas mixture
Secondary phase	Oxygen carrier (F6AL1100)
Average particle diameter	150 μm
Average particle density	2150 kg/m^3
Initial bed mass	~180 g
Fluidizing gas composition in fuel reactor	50% CO, 50% H$_2$
Fluidizing gas composition in air reactor	100% N$_2$
Inlet boundary condition in fuel reactor	Velocity inlet with velocity 0.1 m/s
Inlet boundary condition in air reactor	Velocity inlet with velocity 0.5 m/s
Outlet boundary condition in fuel reactor	Pressure outlet at atmospheric pressure
Outlet boundary condition in air reactor	Pressure outlet at atmospheric pressure
Operating temperature	1123 K
Solids pressure	Lun et al. [7]
Granular bulk viscosity	Lun et al. [7]
Granular viscosity	Gidaspow [8]
Drag law	Gidaspow [8]
Heat transfer coefficient	Gunn [9]
Numerical scheme	Phase-coupled SIMPLE
Time step size	0.0005 s
Iterations per time step	20

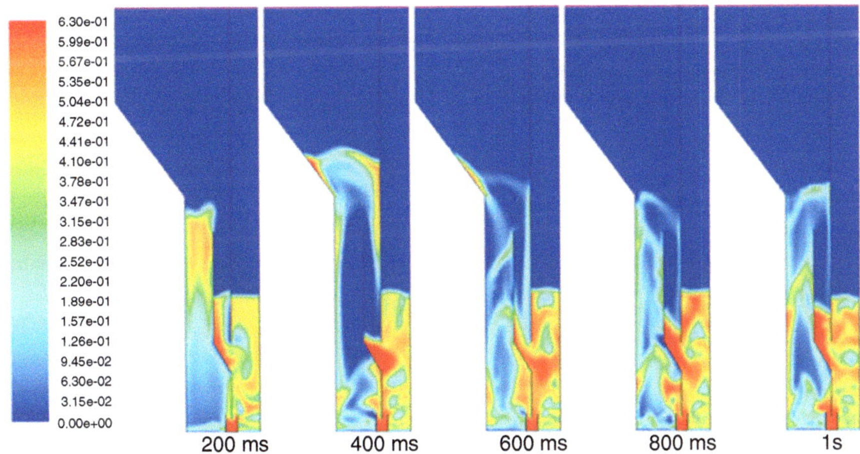

Fig. 5.3 Contours of solid volume fraction for the first second of 2D simulation showing the initial development of solids flow inside the dual-fluidized bed system; the maximum value of 0.63 represents the solids packing limit

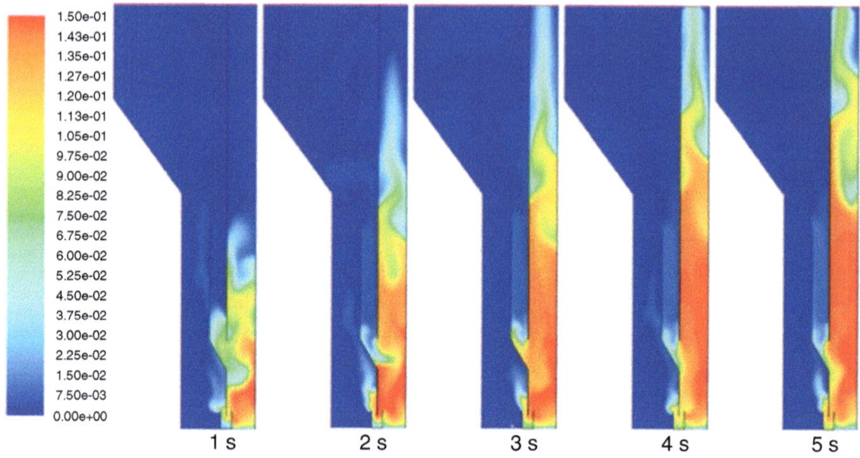

Fig. 5.4 Contours of CO_2 mass fraction for the first 5 s of 2D simulation showing the vortex pattern seen during the initial development of gas flow inside the fuel reactor

in the particle separator leads the particles to drop into the air reactor bed and the downcomer; this leads to a densely packed bed in the downcomer nearing the packing limit. The dense packing is beneficial because it seals the passage and minimizes leakage of nitrogen from the air reactor to the fuel reactor, ensuring that the flue stream of the fuel reactor is a high-purity CO_2 stream uncontaminated by nitrogen ready for capture. It can also be seen that the particles in the fuel reactor are not significantly lifted by the flow because of the lower velocity of the fluidized gas; the bed configuration in the fuel reactor represents a bubbling fluidized bed. Although it is not evident from Fig. 5.3, inspection of the solid phase velocity vectors indicates that the motion of solid particles in the slot or loop seal is in the correct direction, i.e., from the fuel reactor to the air reactor. In this way, the salient features of fluidization behavior in the dual-fluidized bed reactor used by Abad et al. [3] are completely captured by the CFD simulation of the reactor. However, the exact distribution of particles in the experiment could not be observed given the solid walls of the reactor; therefore a direct comparison cannot be made.

The contour plots of the solid volume fraction at subsequent times (not shown) indicate that the flow of solids in the reactor system stabilizes after around 1 s and continues in the same fashion for the remainder of the simulation. On the other hand, the contours of mass fraction of CO_2 given in Fig. 5.4 for the first 5 s of simulation show that the development of the flow in the gas phase takes longer. The simulation is initialized with nitrogen in both reactor beds; this represents the nitrogen cycling performed in the experiment after each reaction period. As the syngas is injected into the fuel reactor, it starts to react with the Fe_2O_3 in the oxygen carrier and reduces it to Fe_3O_4. The CO_2 produced by the reaction forms a plume that first reaches the fuel reactor outlet just after 2 s. As the simulation continues, the plume becomes stronger, as more and more CO_2 is produced by the reduction of Fe_2O_3. It is noted that the mass fraction of H_2O follows the exact same contours as the mass fraction of CO_2

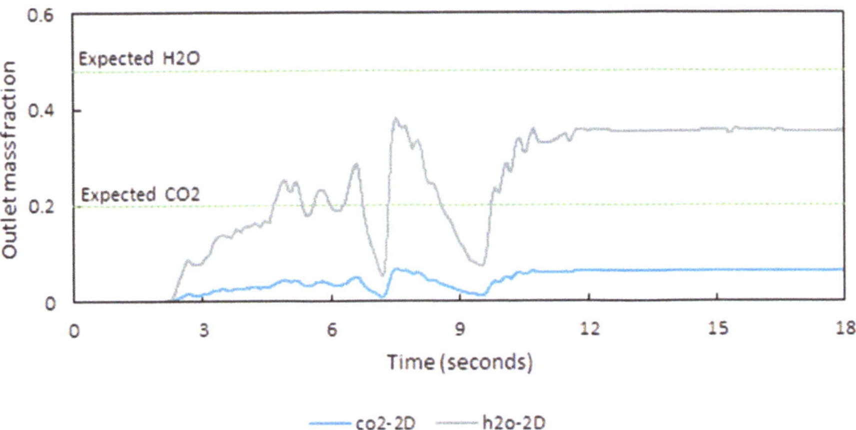

Fig. 5.5 Mass fractions of CO_2 and H_2O at the fuel reactor outlet for the 2D simulation showing fluctuations caused by the vortex pattern seen in Fig. 5.4 during initial development of flow

shown in Fig. 5.4, albeit with different values. Figure 5.4 confirms that the dense packing in the downcomer and slot regions limits the leakage of CO_2 into the air reactor to a very small amount.

An interesting observation from Fig. 5.4 is that the CO_2 plume does not diffuse into the nitrogen present in the reactor from the initialization step; instead, it travels in a vortex pattern toward the fuel reactor outlet. This is because diffusion of gases is an inherently 3D process and cannot be captured accurately in a 2D simulation. Consequently, pockets of reversed flow begin to develop in the fuel reactor. This is reflected in Fig. 5.5, which shows large fluctuations in the concentrations of CO_2 and H_2O measured at the fuel reactor outlet as the pockets of reversed flow lead to alternating CO_2/H_2O-rich and CO_2/H_2O-lean gas at the outlet surface. As the simulation continues, the vortex pattern dissipates and the pockets of reversed flow coalesce into a single, stable plume of reversed flow, which can be clearly seen in Fig. 5.6. Although this stabilizes the concentrations of CO_2 and H_2O at the fuel reactor outlet, their values are much lower than the expected values from Abad et al.'s experiments (indicated in Fig. 5.5). The discrepancy can be explained by the plume of reversed flow consisting of a large amount of nitrogen at the fuel reactor outlet greatly lowering the mass fractions measurements for the other gases.

5.4   Three-Dimensional Simulation of Experiment of Abad et al. [3]

Although the 2D simulation presented in the previous Sect. 5.3 was able to capture the salient features of the fluidization behavior in the dual-fluidized bed system, it was unable to produce the expected concentrations of CO_2 and H_2O in the fuel

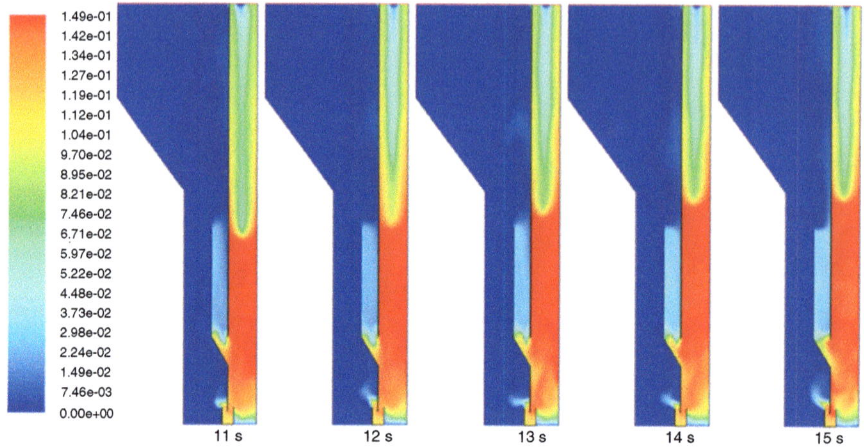

Fig. 5.6 Stable plume of reversed flow in fuel reactor after 10 s of simulation

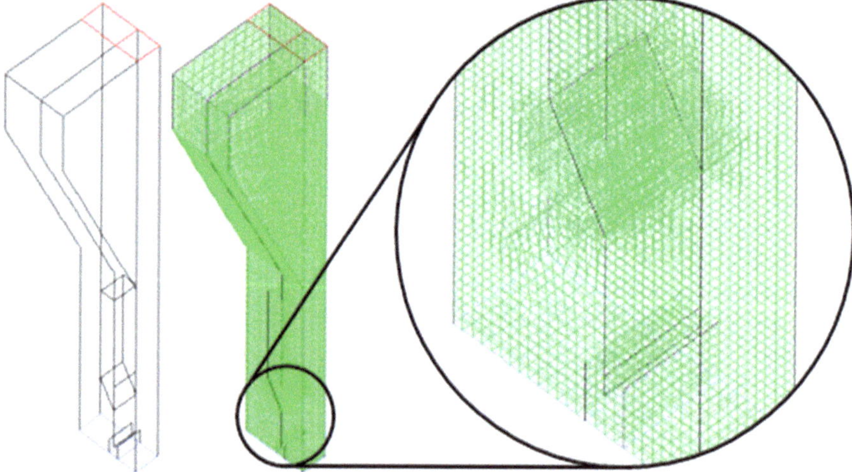

Fig. 5.7 Computational domain (L) and grid (R) for 3D CFD simulation

reactor because of the inadequacy of the 2D simulation in modeling the gaseous diffusion. In this section, a 3D simulation of the experiment of Abad et al. [3] is conducted to obtain a closer fit of the reaction results between the simulation and experiment. Unlike the 2D domain, which used a modified version of the reactor geometry with the particle separator in plane, the 3D computational domain is an exact representation of the geometry in the experiment of Abad et al. [3] Similar to the 2D case, a structured mesh is used with a relatively fine grid in the lower part of the reactor and a coarser grid in the upper regions. The mesh used for the 3D simulations is shown in Fig. 5.7. The initial solids loading in the bed is about

300 g, of which 110 g is in the fuel reactor in line with the experiment. All numerical parameters for the simulation are kept the same as in Table 5.1. The rates for the reactions given in Eqs. (5.1) and (5.2) are identical to those used for the 2D simulation.

The 3D simulation was run for 30 s. Since the solids loading in the current simulation has been increased from the 2D case to match the experimental value, it takes slightly longer to achieve the pressure buildup required for fluidization; the solid particles begin to display fluidization behavior starting at around 5 s. Once the fluidization starts, despite the difference in the configuration of the particle separator (in plane versus out of plane), there are no significant differences in the fluidization behavior of the solids between the 2D and 3D simulations. This is expected because there were no sources of swirl, etc. in the experiment that would lead to a different fluidization behavior in 3D. Indeed, this result validates the reasoning behind conducting a 2D simulation in the first place. Thus, both 2D and 3D simulations can accurately capture all the different fluidization regimes in the experimental dual-fluidized bed reactor of Abad et al. [3]

On the other hand, there are stark differences between the 2D and the 3D simulations when the development of gas flow is investigated. The contours of the mass fraction of CO_2 for the 3D simulation are shown in Fig. 5.8. As expected, there is greater diffusion in the 3D case. The local mass fraction of CO_2 at the base of the bed, where the injected CO first comes into contact with the Fe_2O_3 and begins to react, is around 15%, the same as in the 2D simulation. However, owing to the increased diffusion in 3D, the CO_2 spreads through the fuel reactor more homogeneously as it travels toward the fuel reactor outlet; the vortex patterns seen in the 2D case are absent. The contours of the mass fraction of H_2O display the same characteristics. The quantitative effects of this can be observed from the plot of the

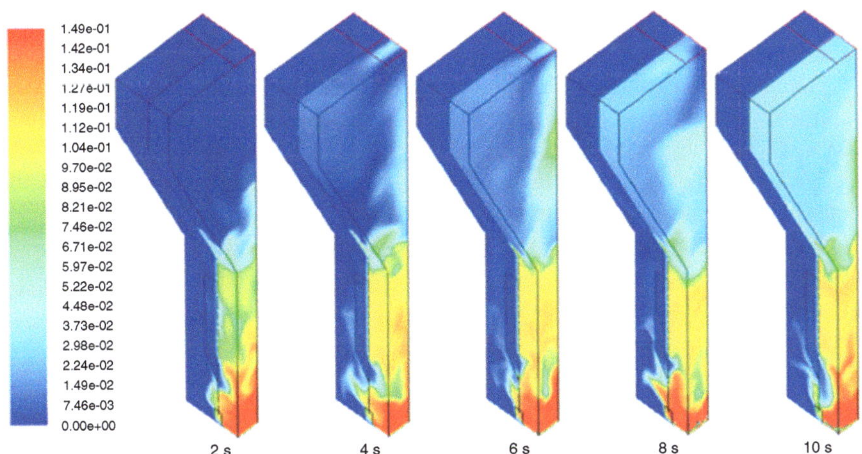

Fig. 5.8 Contours of CO_2 mass fraction for the first 10 s of 3D simulation showing the increased diffusion and absence of the vortex pattern compared to the 2D case

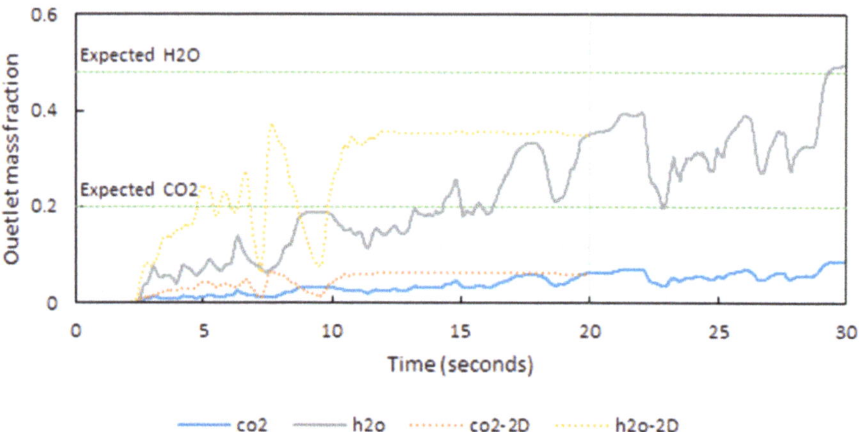

Fig. 5.9 Mass fractions of CO_2 and H_2O at the fuel reactor outlet for 3D simulation

mass fractions of CO_2 and H_2O at the fuel reactor outlet given in Fig. 5.9. The mass fractions of both CO_2 and H_2O are initially lower than in the 2D case. This is because these gases now have to diffuse through the nitrogen present in the fuel reactor instead of displacing it and thus reach the outlet more slowly. The large fluctuations in the outlet mass fraction caused by pockets of reversed flow in the 2D case are eliminated. Since reversed flow does not develop in the current simulation, the mass fractions keep increasing as the simulation progresses and more and more CO_2 and H_2O are produced. By 20 s, the mass fractions of both CO_2 and H_2O have exceeded their stagnation values from the 2D simulation (shown by dotted lines in Fig. 5.9). By 30 s, the mass fraction of H_2O reaches the expected value from the batch experiments of Abad et al. [3] However, it is noted that although the final outlet mass fraction of CO_2 after 30 s is higher than the 2D case, it does not reach the experimental value.

The 3D simulation shows a significant improvement in the mass fraction measurements of the flue gases at the fuel reactor outlet. However, there is still some discrepancy in the mass fraction of CO_2, which may be due to various external factors. In Abad et al.'s experiment [3], the gas streams from the air and fuel reactor outlets were pipelined via an electric cooler into the gas analyzer. It is well established that significant apparent diffusion can occur in gases when they travel through pipes [10]. Thus, it is reasonable to expect that the concentrations measured at the gas analyzer may be different from the concentrations present right at the fuel reactor outlet, which is what is recorded in the simulation. It should also be noted that the reaction rate kinetics used in the simulation were based on the experimental study of Mattisson et al. [5] using pure Fe_2O_3 whereas the oxygen carrier used in the experiment was F6A1100. One of the reasons F6A1100 is preferred over Fe_2O_3 as the oxygen carrier for CLC operation is its improved reactivity [4], caused by an increase in apparent surface area due to the presence of the porous Al_2O_3. Given the improved reactivity, it stands to reason that the experiment shows a higher concentration of the reaction products compared to the simulation. Further research is required to determine more accurate empirical formulas for the reduction of F6A1100 specifically to improve the accuracy of the results of the CFD simulation.

This chapter has provided a typical example of CFD simulation of a fluidized bed reactor using Eulerian-Eulerian two-fluid model. This approach can be applied to other single or dual-fluidized bed reactors.

References

1. Y.-O. Chong, D.J. Nicklin, P.J. Tait, Solids exchange between adjacent fluid beds without gas mixing. Powder Technol. **47**(2), 151–156 (1986)
2. M. Fang, C. Yu, Z. Shi, Q. Wang, Z. Luo, K. Cen, Experimental research on solid circulation in a twin fluidized bed system. Chem. Eng. J. **94**(3), 171–178 (2003)
3. A. Abad, T. Mattisson, A. Lyngfelt, M. Johansson, The use of iron oxide as oxygen carrier in a chemical-looping reactor. Fuel **86**(7–8), 1021–1035 (2007)
4. M. Johansson, *Screening of Oxygen-Carrier Particles Based on Iron-, Manganese-, Copper- and Nickel Oxides for Use in Chemical-Looping Technologies* (Chalmers University of Technology, Göteborg, 2007)
5. T. Mattisson, J.C. Abanades, A. Lyngfelt, A. Abad, M. Johansson, J. Adanez, F. Garcia-Labiano, L.F. de Diegos, P. Gayan, B. Kronberger, *Capture of CO2 in Coal Combustion. ECSC Coal RTD Programme Final Report*, Rep. No. ECSC-7220-PR125 2005 (Chalmers University of Technology, Göteborg)
6. K. Mahalatkar, J. Kuhlman, E.D. Huckaby, T. O'Brien, CFD simulation of a chemical-looping fuel reactor utilizing solid fuel. Chem. Eng. Sci. **66**(16), 3617–3627 (2011)
7. C.K.K. Lun, S.B. Savage, D.J. Jeffrey, N. Chepurniy, Kinetic theories for granular flow: Inelastic particles in Couette flow and slightly inelastic particles in a general flowfield. J. Fluid Mech. **140**, 223–256 (1984)
8. D. Gidaspow, D. Gidaspow, Multiphase flow and fluidization: Continuum and kinetic theory descriptions. AICHE J. **42**(4), 1197–1198 (1994)
9. D.J. Gunn, Transfer of heat or mass to particles in fixed and fluidised beds. Int. J. Heat Mass Transf. **21**(4), 467–476 (1978)
10. G.I. Taylor, The dispersion of matter in turbulent flow through a pipe. Proc. R. Soc. Lond. Ser. A Math. Phys. Sci. **223**(1155), 446–468 (1954)

Chapter 6
Lagrangian Simulation of a CLC Reactor

6.1 Simulation of Spouted Fluidized Bed Using Fe_2O_3 as Bed Material

Link [1] identified different gas-solid flow regimes in multiphase flow in a fluidized bed with relatively large diameter particles of glass beads with a density of 2500 kg/m^3. In recent years, Lagrangian multiphase simulations have captured the particle dynamics in spouted fluidized beds for particle densities ranging from 1500 to 2500 kg/m^3 [2–4]. In the simulation example considered here, chemical reactions are incorporated into the CFD-DEM model of a spouted fluidized bed reactor of Zhang et al. [4] while employing particles of Fe_2O_3 with a density of 5240 kg/m^3 as the bed material. As such, the results of this simulation provide an understanding of the effect of material density on the performance of the spouted fluidized bed reactor.

The geometry and computational model of the CLC reactor are shown in Fig. 6.1. The geometry is derived from the pseudo-2D Plexiglas test rig used in the TU-Darmstadt cold-flow experiment [3] with the chute structure added in the simulation to improve particle circulation based on the work of Zhang et al. [4] The mesh is generated such that the solution is stable when using the second-order numerical schemes and minimal under-relaxation to achieve faster convergence at each time step. The particle diameter is kept constant at 2.5 mm; any smaller sized of particles makes the total number of particles required to maintain a reasonable bed height prohibitively large for individual particle tracking in the computation. However, the total particle load in the bed is approximately doubled compared to the work of Zhang et al. [4] to partly offset the tighter packing associated with the heavier particles. Additional particles are also deposited in the downcomer and loop seal to ensure adequacy of particles for recirculation; a total of 87,320 particles are employed in the entire system.

According to Link [1], the flow regime inside the fluidized bed depends on the ratio of the spout jet velocity and the background velocity to the minimum fluidization velocity of the particles; in order to achieve a spouted fluidized bed, these

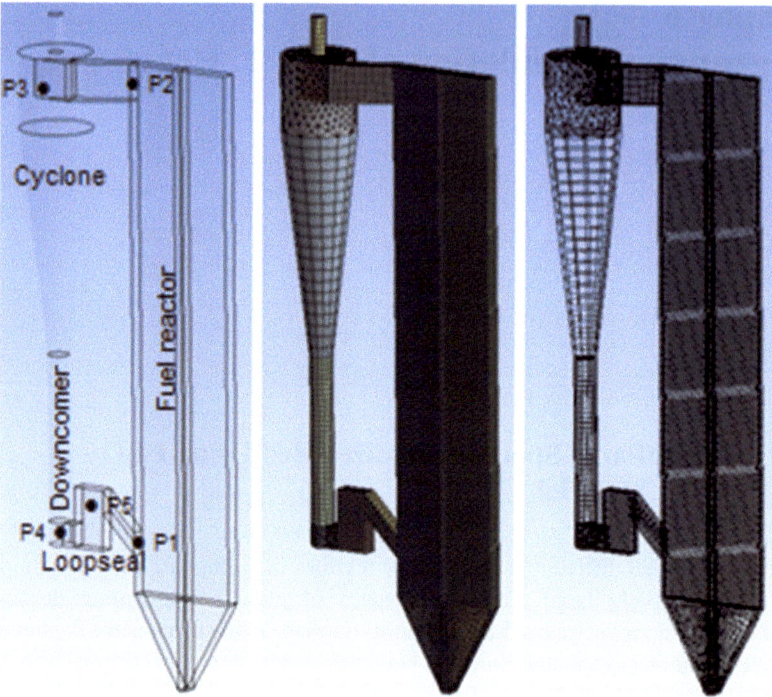

Fig. 6.1 Geometry outline with pressure taps, mesh, and wireframe of the CLC reactor

velocities are increased in the simulation reported here compared to the velocities used by Zhang et al. [4] employing particle of 2500 kg/m^3 density in order to compensate for the higher minimum fluidization velocity of the heavier Fe_2O_3 particles. The velocities are also increased to counteract the reduced momentum transfer from the gas phase to the solid phase in the presence of chemical reactions. Thus, the central jet velocity is increased to 50 m/s, and the background velocity is increased to 2 m/s.

The flow injection in the fuel reactor consists of 10% CH_4 and 10% H_2O by mass fraction. The reaction kinetics for the reduction of Fe_2O_3 to Fe_3O_4 by the CH_4 are based on the experimental work of Son and Kim [5]. The remaining 80% of the flow injection is inert nitrogen. The absence of CO_2 in the fluidization gas is maintained so that the mass fraction of the CO_2 generated by the reaction between Fe_2O_3 and CH_4 can be tracked without it being overshadowed by the injected mass fraction of CO_2. Similarly, the aeration gas in the downcomer and the loop seal comprises solely of N_2 so that the recirculation of particles from the loop seal to the fuel reactor can be easily identified; the particles that originate in the loop seal have a smaller mass fraction of Fe_3O_4 since they were initially exposed to inert flow. All other parameters in the simulation are kept unchanged from the cold-flow simulation work of Zhang et al. [4] The key numerical parameters in the present simulation are summarized in Table 6.1.

Table 6.1 Key modeling parameters for reacting flow simulation in the CLC fuel reactor

Average diameter of particles	2.5 mm
Average density of particles	5240 kg/m^3
Mass load of particle in bed	~1.3 kg
Primary phase	Gaseous mixture of CH$_4$, H$_2$O, CO$_2$, N$_2$
Secondary phase	Solid mixture of Fe$_2$O$_3$, Fe$_3$O$_4$
Outlet boundary condition	Pressure outlet with $P_{out,gage} = 0$ Pa
Inlet boundary condition	Velocity inlet with central jet velocity of 50 m/s, background flow velocity of 2 m/s
Drag law	Syamlal-O'Brien
Particle collision law	Spring-dashpot
Spring constant	410 kN/m
Coefficient of restitution	0.97
Friction coefficient	0.5
Numerical schemes	Phase coupled SIMPLE, second-order upwind for momentum equation, QUICK for volume fraction, second-order upwind for energy, second-order upwind for species; second-order implicit in time
Time-step size	Particle: 1×10^{-5} s, fluid: 1×10^{-4} s

The simulation is carried out on a Dell workstation using a six-core Intel Xenon CPU. Each run requires about 24 hours of CPU time per 200 ms of simulation time. The particle distributions and velocities are inspected at 40 ms intervals and presented in Fig. 6.2.

The particle tracks in Fig. 6.2 show that a prominent bubble forms almost immediately as a result of the initial pressure build-up in the fuel reactor. However, owing to the increased particle mass, the velocity of the particles is not sufficient to carry the particles all the way to the top of the reactor and into the cyclone. The bubble bursts around 520 ms as evidenced by the velocity of the particles at the top of the bubble reaching zero, and the particles start to fall back into the fuel reactor bed. Moreover, despite the increased total number, the particles initially deposited into the downcomer and the loop seal become tightly packed at the bottom because of the increased particle mass. As a result, there are not enough particles available to recirculate back into the fuel reactor.

Figure 6.2 indicates that the performance of the spouted fluidized bed has a strong dependence on the bed mass, which, in turn, depends on the density of the bed material. With the density of Fe$_2$O$_3$ of 5240 kg/m^3, the bubble formed in the fuel reactor does not have sufficient energy to reach the top of the reactor. Increasing the central jet velocity further may prevent this issue, but it leads to the formation of a straight pathway through the dense bed region once the initial pressure build-up is lost, which prevents the critical pressure build-up required for subsequent bubbles.

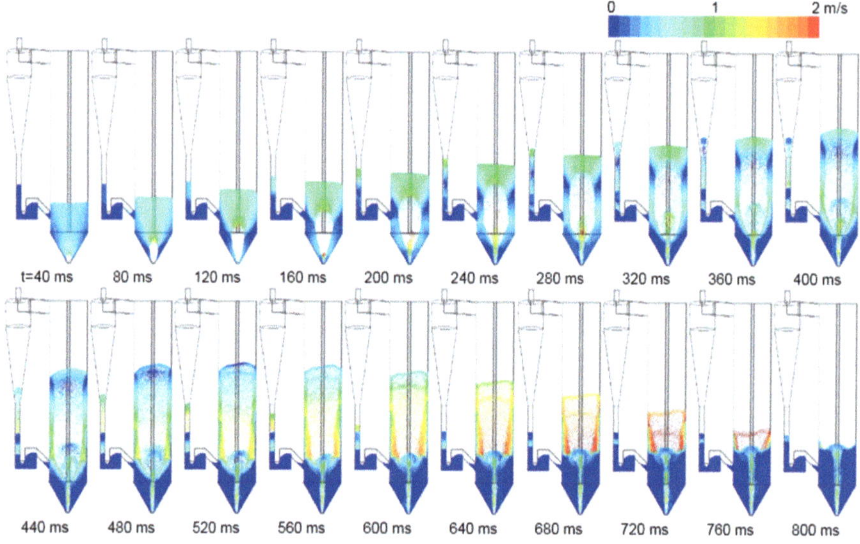

Fig. 6.2 Particle tracks colored by velocity magnitude in reacting flow with Fe_2O_3 particles

Such a pathway is already visible to a certain extent in Fig. 6.2 from around 680 ms onward. Another alternative is to reduce the height of the reactor such that the reduced momentum transferred to the particles is still sufficient for the particles to reach the top of the reactor and into the cyclone. However, this also reduces the residence time of the fuel and affects the fuel conversion rate, particularly using solid coal fuel, which has to go through a slow gasification process before it can react with the metal oxide. Since the particles' tracks in Fig. 6.2 clearly demonstrate that the spouted fluidized bed setup employed here is a poor choice, any additional results and the discussion on the chemical reactions are deferred to the next section in the interest of conciseness.

6.2 Simulation of Spouted Fluidized Bed Using Fe_2O_3 Supported on $MgAl_2O_4$ as Bed Material

In order to address the underlying problem of poor fluidization associated with the high-density Fe_2O_3 particles, it has been proposed to use an oxygen carrier consisting of an active part based on Fe combined with an inert support material. The support material is porous and serves to increase the reaction kinetics by providing a larger surface area for reaction as well as to reduce the overall density of the particle. Suitable candidates are $MgAl_2O_4$ and SiO_2, among others [6]. Of these, the work of Johansson et al. [7] showed that the oxygen carrier consisting of 60% Fe_2O_3 by mass on 40% $MgAl_2O_4$ support sintered at 1100 °C, designated F60AM1100, displayed excellent reactivity and sufficient hardness, and its apparent

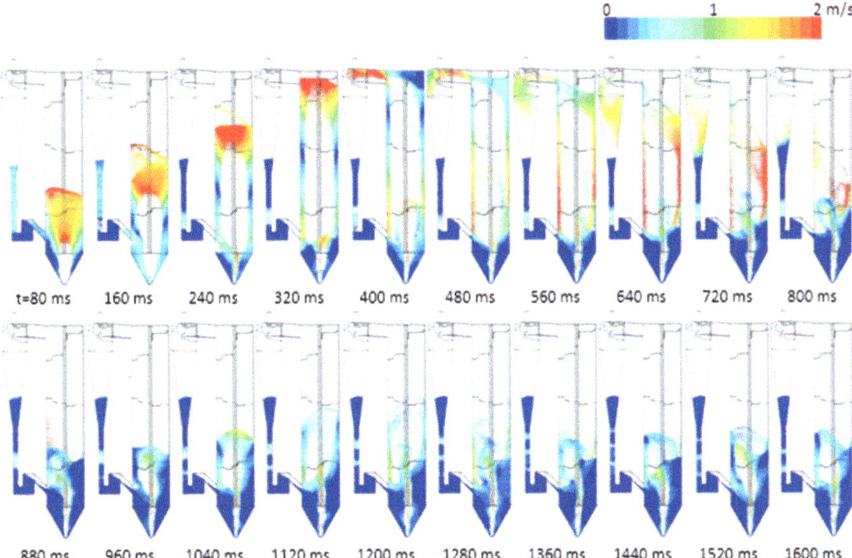

Fig. 6.3 Particle tracks colored by the velocity magnitude in reacting flow with F60AM1100 particles

density of 2225 kg/m^3 makes it an ideal choice for fluidized bed operation. Here, the performance of a spouted fluidized bed reactor with chemical reactions is provided and discussed using the F60AM1100 oxygen carrier as the bed material.

Since MgAl$_2$O$_4$ is inert, the only reaction that takes place on the particle surface is the reduction of Fe$_2$O$_3$ to Fe$_3$O$_4$ by the injected CH$_4$ fuel. The stoichiometric reaction is given by

$$12Fe_2O_3(s) + CH_4 \rightarrow 8Fe_3O_4(s) + CO_2 + 2H_2O \qquad (6.1)$$

Because of the lower density of F60AM1100, the fluidization velocity in the fuel reactor is reduced to 40 m/s, and the total particle load is around 0.7 kg. All other parameters in this simulation remain unchanged from Table 6.1. The computational cost for this setup is similar to that described for the previous setup. Figure 6.3 shows the particle distributions and velocities at 80 ms intervals for the first 1600 ms of flow injection.

Figure 6.3 shows that a prominent gas bubble forms in the CLC reactor from 0 to around 400 ms. The leading front of the spout reaches the top of the fuel reactor around 400 ms, and a large number of particles are deposited into the cyclone through the connecting duct. Between around 480 and 880 ms, the pressure build-up in the fuel reactor vanishes, and the remaining particles in the fuel reactor fall back into the fluidized bed while the particles in the cyclone fall into the downcomer. Some recirculation of the particles from the loop seal back into the fuel reactor is also evident. Once the particles start to settle back into the fluidized bed, aided by the

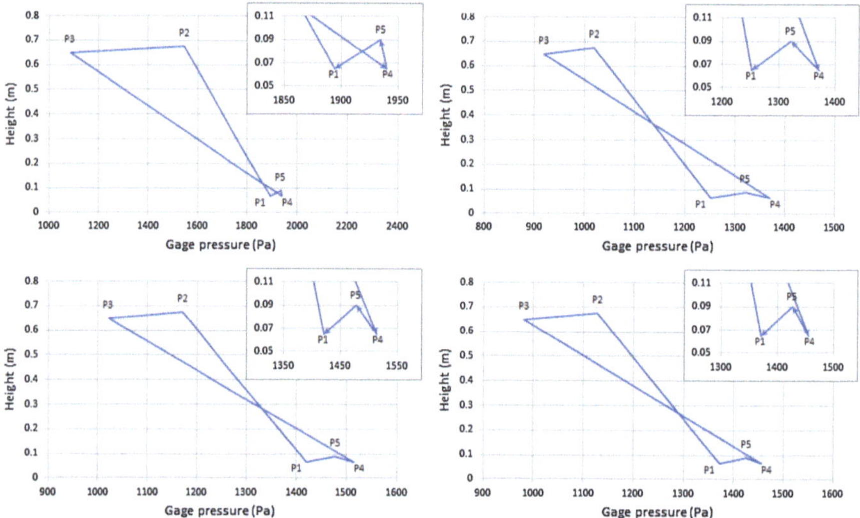

Fig. 6.4 Static pressure at pressure taps P1–P5 in the CLC system of Fig. 6.1 at $t = 400$ ms (top left), 800 ms (top right), 1200 ms (bottom left), and 1600 ms (bottom right) in reacting flow

recirculation of particles from the loop seal, the pressure build-up due to the jet injection is partially restored and subsequent gas bubbles are formed around 960 and 1440 ms. However, in these cases, the kinetic energy transferred to the particles is insufficient to carry them to the top of the reactor, and the bubbles collapse prematurely. This can be explained by the bypass pathway formed at the same time by the high-velocity jet in the absence of the initial packed bed, which allows the energy in the jet to bypass the dense bed region and prevents the critical pressure build-up in the fuel reactor. Since the gas bubble formation and the particle recirculation are both driven by the pressure at various locations in the system, the static pressure readings at pressure taps P1 through P5 are investigated to better understand the behavior observed from the particle tracks in Fig. 6.3; the pressure tap data is presented in Fig. 6.4.

The subplot at 400 ms shows a large pressure build-up of around 1900 Pa at P1 (base of the reactor). At 800 ms, when the bubble has collapsed, the initial pressure build-up is lost as the pressure at P1 drops to around 1250 Pa. Subsequently, at 1200 and 1600 ms, the pressure at P1 increases slightly to around 1400 and 1350 Pa, respectively. This build-up of pressure is in line with the observation of the second and third gas bubble formations; the slight increase compared to the initial bubble also explains why the subsequent bubbles did not carry sufficient kinetic energy to reach the top of the reactor. From Fig. 6.4, it can also be noted that there is a consistent positive pressure differential between taps P4 (base of the loop seal) and P1 of around 100 Pa, which corroborates the continuous recirculation of particles from the loop seal back into the fuel reactor observed from the particle tracks.

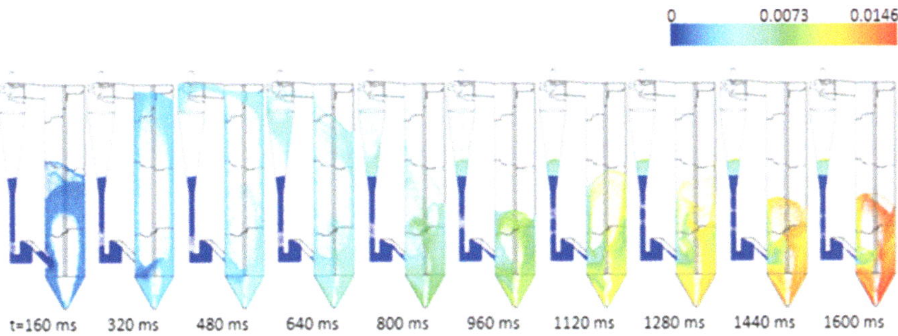

Fig. 6.5 Particle tracks colored by mass fraction of Fe$_3$O$_4$ relative to original mass of Fe$_2$O$_3$

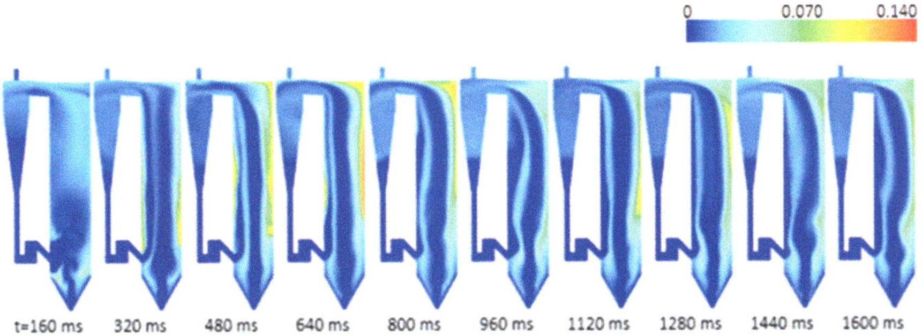

Fig. 6.6 Contours of CO$_2$ mass fraction produced by reaction of Fe$_2$O$_3$ with CH$_4$

Based on the diameter and density, the F60AM1000 particles used as the oxygen carrier can be classified as Group D particles according to Geldart's powder classification [8]. In the absence of experimental results of spouted fluidized bed operation of reacting flow with Group D particles, the successful incorporation of chemical reactions into the multiphase flow simulation is judged by inspecting the formation of Fe$_3$O$_4$ and CO$_2$ as a result of the reaction. These results are presented in Figs. 6.5 and 6.6, respectively.

From Fig. 6.5, it can be seen that the mass fraction of Fe$_3$O$_4$ increases with time for the particles inside the fuel reactor as the simulation advances as expected. The continuous particle recirculation is also evident from the consistent presence of a small number of particles in the fuel reactor with a lower mass fraction of Fe$_3$O$_4$ since these particles originated in the loop-seal region where the flow is inert. According to Fig. 6.6, the mass fraction of CO$_2$ rises quickly in the first 640 ms of simulation, after which it drops slightly before spiking again around 1120 ms. Such fluctuations are due to the inherently unsteady nature of the solid-gas mixing in the fuel reactor. Overall, the observations from Figs. 6.5 and 6.6 indicate successful incorporation of chemical reactions into the CFD/DEM model for the CLC reactor configuration developed by Zhang et al. [4]

Comparing Fig. 6.3 with Fig. 6.2, it is clear that changing the bed material in the fuel reactor from Fe_2O_3 to F60AM1100 consisting of 60% Fe_2O_3 and 40% $MgAl_2O_4$ significantly improved the fluidization performance of the reactor. In the current simulation, the inadequacy of the central jet to impart sufficient momentum to the particles for them to reach the top of the reactor was rectified, and continuous particle recirculation was observed through the loop seal. However, the formation of the bypass pathway through the bed after the first bubble collapses remains a concern and hinders the pressure build-up required for subsequent bubbles to reach the top of the reactor. Since F60AM1100 is already among the lightest Fe-based oxygen carriers studied by Johansson et al. [7] and lighter alternatives likely to be expensive, one alternate way to mitigate this problem is to use a cyclic flow injection whereby the jet is turned off intermittently to allow the bed particles to resettle down into the original packed bed configuration, which resets the fluidization behavior to the initial bubble formation stage once the jet is turned back on. Cyclic injections have already been used in laboratory-scale CLC experiments such as in the work of Son and Kim [5] to switch between N_2 and CH_4 in lieu of separate fuel and air reactors, and their operational feasibility in an industrial setting can be readily studied.

6.3 Simulation of Spouted Fluidized Bed with Pseudo-coal Injection

The spouted fluidized bed was proposed as a viable configuration for CLC with direct coal injection. The addition of coal particles into the fluidized bed system investigated in Sects. 6.2 and 6.3 would convert the system into a binary particle bed, and the collisions between the oxygen carrier and coal would require special attention. In this section, only the reaction mechanisms for solid coal are investigated with a pseudo-coal injection that represents the gaseous products of coal after the devolatilization and gasification steps occur. The pseudo-2D reactor in the experiment at TU-Darmstadt used in the reacting flow simulations in the previous section is considered again with the geometry and mesh shown in Fig. 6.1.

The work of Merrick [9] provides a methodology for predicting the compositions of char, tar, and volatile species based on the physical and chemical properties of the coal. The proximate and ultimate analysis of the South African coal considered in the pseudo-coal simulation is given in Table 6.2.

Table 6.2 Physical and chemical properties of South African coal

	Proximate analysis (wt.%)				Ultimate analysis (wt.%)				
	Moisture	Volatile matter	Fixed carbon	Ash	C	H	N	S	O
South African coal	8.3	21.6	54.2	15.9	62.5	3.5	1.4	0.7	7.7
Dry ash-free basis	–	28.5	71.5	–	85.5	4.6	1.8	0.9	10.1

Table 6.3 Constituent mass fractions of char and tar species based on the experiment of Merrick [9]

	C	H	O	N	S
Char (coke)	0.98	0.002	0.002	0.01	0.006
Tar	0.85	0.082	0.049	0.009	0.006

The approach suggested by Merrick [9] to predict the final yields of the volatile matter species is to construct a set of simultaneous linear equations written as $A_{ij}m_j = b_i$ where m_j is the vector of unknowns representing the yields of char (coke), CH_4, C_2H_6, CO, CO_2, tar, H_2, H_2O, NH_3, and H_2S as mass fractions of the daf coal, and A_{ij} and b_j are matrix and vector of constants, respectively. The list of species considered by Merrick [9] is not exhaustive; additional volatile matter species could be modeled if suitable data were available. The first five equations represent element balances on carbon, hydrogen, oxygen, nitrogen, and sulfur, respectively. The corresponding values in A_{ij} (for $i = 1,\ldots, 5$ and $j = 1,\ldots, 10$) represent the analyses of the volatile matter species expressed as the mass fractions of the respective species in b_j obtained from the ultimate analysis of the daf coal. The char and tar species are assumed to have a specific composition in terms of the constituent elements given in Table 6.3.

The sixth equation describes the yield of char as a function of the total volatile matter release obtained from the proximate analysis. Merrick [9] obtained a correlation for predicting the volatiles released by coal as

$$V = p - 0.36\,p^2 \tag{6.2}$$

where p is the volatile matter from the proximate analysis in the daf coal. For the South African coal considered in the current study with $p = 0.285$, the total volatile released is $V = 0.256$. The remaining four equations provide the yields of the remaining volatile species in terms of the ultimate analysis based on approximate evolutions of the hydrogen and oxygen species in the coal in the final yield of volatile matter. Merrick [9] found that the yield of CH_4 and C_2H_6 accounted for 32.7% and 4.4% of the hydrogen in the coal, respectively, and that 18.5% and 11.0% of the oxygen in the coal evolved in the CO and CO_2 species. Based on these assumptions, the final set of simultaneous equations to obtain the total volatile matter yields can be written as

$$
\begin{bmatrix}
0.98 & 0.75 & 0.8 & 0.4286 & 0.2727 & 0.85 & 0 & 0 & 0 & 0 \\
0.002 & 0.25 & 0.2 & 0 & 0 & 0.082 & 1 & 0.1111 & 0.1765 & 0.0588 \\
0.002 & 0 & 0 & 0.5714 & 0.7273 & 0.049 & 0 & 0.8889 & 0 & 0 \\
0.01 & 0 & 0 & 0 & 0 & 0.009 & 0 & 0 & 0.8235 & 0 \\
0.006 & 0 & 0 & 0 & 0 & 0.01 & 0 & 0 & 0 & 0.9412 \\
1 & 0 & 0 & 0 & 0 & 0 & 0 & 0 & 0 & 0 \\
0 & 1 & 0 & 0 & 0 & 0 & 0 & 0 & 0 & 0 \\
0 & 0 & 1 & 0 & 0 & 0 & 0 & 0 & 0 & 0 \\
0 & 0 & 0 & 1 & 0 & 0 & 0 & 0 & 0 & 0 \\
0 & 0 & 0 & 0 & 1 & 0 & 0 & 0 & 0 & 0
\end{bmatrix}
\begin{Bmatrix} \text{Char} \\ CH_4 \\ C_2H_6 \\ CO \\ CO_2 \\ \text{Tar} \\ H_2 \\ H_2O \\ NH_3 \\ H_2S \end{Bmatrix}
=
\begin{Bmatrix} C \\ H \\ O \\ N \\ S \\ 1{-}V \\ 1.31\,H \\ 0.22\,H \\ 0.32\,O \\ 0.15\,O \end{Bmatrix}
\tag{6.3}
$$

The final yields of the volatile matter and char can be determined by calculating $\boldsymbol{m}_j = A_{ij}^{-1}\boldsymbol{b}_i$ and can be written in terms of the mass fractions as

$$1 \ \text{kg daf coal} \rightarrow 0.744\,\text{char} + 0.0277\,\text{tar} + 0.0605\,CH_4 + 0.0102\,C_2H_6$$
$$+ 0.0325\,CO + 0.0152\,CO_2 + 0.0140\,H_2 + 0.0777\,H_2O \qquad (6.4)$$
$$+ 0.0131\,NH_3 + 0.00477\,H_2S$$

Considering the base South African coal with ash and moisture,

$$1\,\text{kg coal} \rightarrow 0.758\,\text{daf char} + 0.159\,\text{ash} + 0.083\,\text{moisture} \rightarrow 0.744\,\text{char}$$
$$+ 0.0277\,\text{tar} + 0.0605\,CH_4 + 0.0102\,C_2H_6 + 0.0325\,CO + 0.0152\,CO_2$$
$$+ 0.0140\,H_2 + 0.0777\,H_2O + 0.0131\,NH_3 + 0.00477\,H_2S$$

$$(6.5)$$

The mass balance in Eq. (6.5) can be converted into a mole balance by considering the molecular weight of each species based on its constituent elements. The molecular weight of ash is assumed to be 100 kg/kmol. The final yield from primary devolatilization of 1 kmol of South African coal is given by

$$1\,\text{kmol coal} \rightarrow 0.684\,\text{char} + 0.0221\,\text{tar} + 0.0426\,CH_4 + 0.00381\,C_2H_6$$
$$+ 0.0131\,CO + 0.00390\,CO_2 + 0.0790\,H_2 + 0.0486\,H_2O$$
$$+ 0.00867\,NH_3 + 0.00158\,H_2S + 0.0236\,\text{ash} + 0.0685\,\text{moisture}$$

$$(6.6)$$

Before the gaseous injection prescribing the devolatilization products can be implemented into the reacting flow simulation, the fate of the solid char and tar species must be resolved. This is done by considering the gasification of char by steam given by the stoichiometric relation

$$\text{Char} \rightarrow 2\,CO + 0.842\,H_2 + 0.00153\,H_2O + 0.00438\,N_2 + 0.00230\,H_2S \quad (6.7)$$

and the secondary devolatilization of tar determined by Bradley et al. [10] assuming a molar ratio of CH_4 to H_2 of 0.5 similar to the primary devolatilization step written as

$$\text{Tar} \rightarrow 0.805\,C_{\text{soot}} + 0.142\,CH_4 + 0.0432\,CO + 0.285\,H_2 + 0.00907\,HCN$$
$$+ 0.00441\,H_2S \qquad (6.8)$$

Hence, the final molar balance after the complete devolatilization and gasification steps is given by

$$1\,\text{kmol coal} \rightarrow 0.0105\,C_{\text{soot}} + 0.0269\,CH_4 + 0.00224\,C_2H_6 + 0.814\,CO$$
$$+0.00230\,CO_2 + 0.0536\,H_2 + 0.0292\,H_2O + 0.00510\,NH_3 \quad (6.9)$$
$$+0.00191\,H_2S + 0.0139\,\text{ash} + 0.0403\,\text{moisture}$$

The product species in Eq. (6.9) are used to specify the pseudo-coal injection for the reacting flow simulation of coal with the Fe-based oxygen carrier F60AM1100 consisting of 60% Fe_2O_3 by mass on inert $MgAl_2O_4$ support. The injection mole flow rate is set at 0.119 kmol/s of coal to represent a mass injection rate of 1.2 kg/s with a central jet velocity of 30 m/s. The mole fractions of the reacting species, namely, CH_4, C_2H_6, CO, and H_2, are kept intact. The remaining species in Eq. (6.9) are collected into a single inert injection of N_2 for simplicity.

Since the devolatilization and gasification of the coal are already accounted for by the pseudo-coal injection, the only reaction mechanisms implemented in the simulation are for the reduction of Fe_2O_3 by CH_4, C_2H_6, CO, and H_2 to form Fe_3O_4, given as

$$12\,Fe_2O_3 + CH_4 \rightarrow 8\,Fe_3O_4 + 2H_2O + CO_2 \quad (6.10)$$

$$21\,Fe_2O_3 + C_2H_6 \rightarrow 14\,Fe_3O_4 + 3H_2O + 2CO_2 \quad (6.11)$$

$$3\,Fe_2O_3 + CO \rightarrow 2\,Fe_3O_4 + CO_2 \quad (6.12)$$

$$3\,Fe_2O_3 + H_2 \rightarrow 2\,Fe_3O_4 + H_2O \quad (6.13)$$

The reactions are incorporated into the CFD-DEM simulation using the particle surface reactions model outlined in this section. The reaction rates for the metal oxide reduction reactions follow the Arrhenius rate equation $k = k_0 \exp(-E_a/RT)$ where k is the reaction rate, k_0 is the pre-exponent factor, E_a is the activation energy, R is the universal gas constant ($= 8.314$ J/K/mol), and T is the temperature of the fluid phase. The values of k_0 and E_a are obtained from Mahalatkar et al. [11] and are summarized in Table 6.4. It should be noted that the reaction rate for the reduction with C_2H_6 is assumed to be the same as that with CH_4 because of the lack of experimental data.

The particle tracks colored by velocity magnitude are shown in Fig. 6.7. An initial bubble forms that carries the bulk of the particles to the top of the fuel reactor and into the cyclone around 400 ms, while the remaining particles fall back into the bed. Once the particles settle back down in the bed, a second bubble starts to develop around 900 ms. However, the formation of a bypass pathway is again evident as in Fig. 6.3, and gauging by the particle velocities, it is expected that the second bubble will not reach the top of the reactor.

Table 6.4 Pre-exponent factor and activation energy for the metal oxide reduction reactions	Reducing agent	k_0 (1/s)	E_a (kJ/mol)
	CH_4	5.33×10^{-4}	24.0
	CO	6.20×10^{-4}	20.0
	H_2	2.30×10^{-3}	24.0

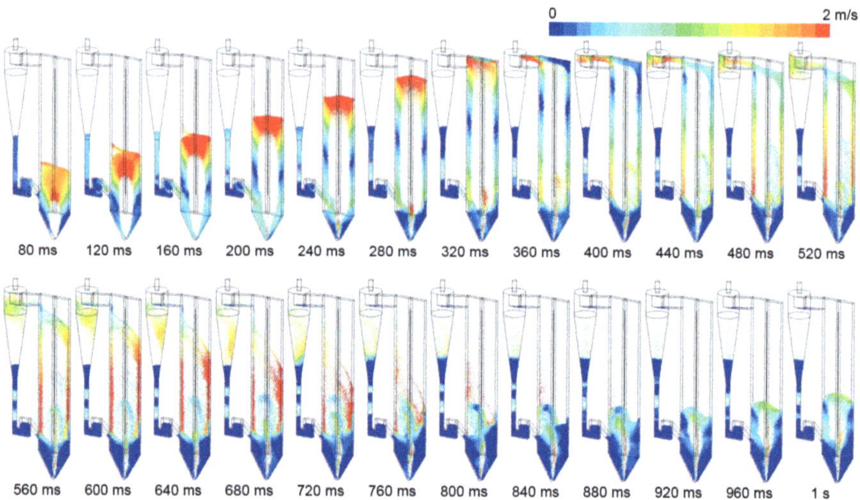

Fig. 6.7 Particle tracks colored by velocity magnitude of F60AM1100 particles in reacting flow with a gaseous injection representing the products of devolatilization and gasification of coal

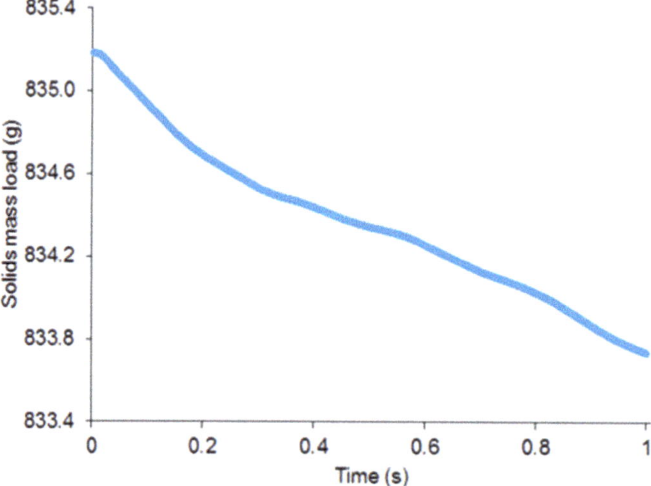

Fig. 6.8 Change in solids mass load in the CLC fuel reactor due to the reduction of F60AM1100 particles by the gaseous products of coal devolatilization and gasification

Since the flow conditions and physical properties of the solid and gas phases are unchanged, the fluidization behavior is similar as expected. The results of the reaction mechanisms with coal provide greater insight. One measure of the progress of the reaction is the change in the solids mass load in the fuel reactor as shown in Fig. 6.8. As the active part of the oxygen carrier, Fe_2O_3, is reduced to Fe_3O_4, the total solid mass reduces. The rate of change suggests that the reaction starts out fast and

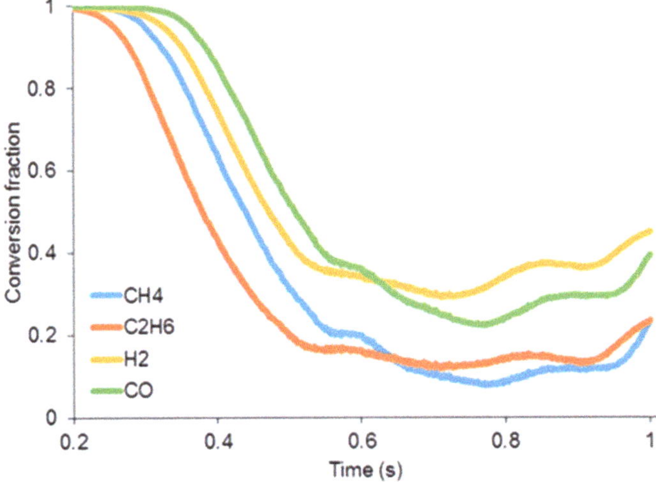

Fig. 6.9 Conversion fraction of the gasifying agents in the CLC fuel reactor due to the reduction of F60AM1100 particles by the gaseous products of coal devolatilization and gasification

starts to decrease as time advances as the surface area of Fe_2O_3 available for reaction reduces due to the formation of Fe_3O_4. A similar trend was observed in the work of Mahalatkar et al. [11] using the Eulerian multi-fluid approach to model the CLC reactor of Leion et al. [12] using solid coal.

The conversion fraction of the reacting species, namely, CH_4, C_2H_6, H_2, and CO, is calculated as a ratio of the outlet molar flow rate from the injection flow rate subtracted from unity and is shown in Fig. 6.9. It should be noted that the initial conversion fraction of 1 for each species is a result of the gases not yet reaching the outlet at the start of the simulation as it takes time to travel through the fuel reactor and does not represent a complete depletion of the species. From around 0.6 s onward, the conversion fraction represents actual values due to species depletion. Given that the reduction of H_2 has the highest value of the pre-exponent factor k_0 (see Table 6.4), it makes sense that H_2 has the highest conversion fraction approaching 0.4, followed by CO due to the lower activation energy compared to CH_4. Experimental data of a spouted fluidized bed system for CLC with reacting flow is not available in the literature. However, the conversion fractions obtained in this simulation are in line with CLC systems using a bubbling or fast fluidized bed with smaller oxygen carrier particles and suggest that the spouted fluidized bed is a viable configuration for CLC once the formation of the bypass pathway can be addressed.

6.4 Simulation of Moving Bed Air Reactor

Moving bed technology has been widely used in industrial process because of the advantages of low-pressure drop, stable operating condition, and sufficient gas-solid contact. In the CLC system, a tower-type moving bed can be used as the air reactor [13]. The flow characteristics in the dense granular system of the moving bed are investigated in this section using the CFD-DEM method [14].

To reduce the computational cost, the original multistage tower-type air reactor is simplified to one stage, and a quasi-2D model is used to describe the simplified air reactor. The depth of the geometry in z direction is about five times the particle diameter. The size of the original geometry is also reduced. Figure 6.10 displays the geometry, its dimensions, and mesh in this simulation. It can be seen that a loop seal, a wedge, and a downcomer are connected with the simplified air reactor, which is to realize the flexible control of the solid flow rate. Cartesian grids with 6315 cells are used to discretize the geometry. The computational cells have a width of 1.8 mm, a height of 1.8 mm, and a depth of 0.8 mm.

Initially, particles with density of 2600 kg/m^3 and diameter of 0.45 mm are packed in the moving bed, wedge, downcomer, and horizontal pipe of the loop seal. The total number of particles is about 116,700. The top of the moving bed is set as the solid inlet. The solid outlet is located at the right of the inclined pipe of the loop seal, while the gas inlet is located at the bottom of the vertical pipe. The front and back planes of the geometry are set as the free-slip wall boundary conditions for the gas phase to decrease their influence in the solid particles' movement, which can make the quasi-2D model closer to a 3D model. The left and right sides are set as the no-slip wall boundary conditions for the gas phase.

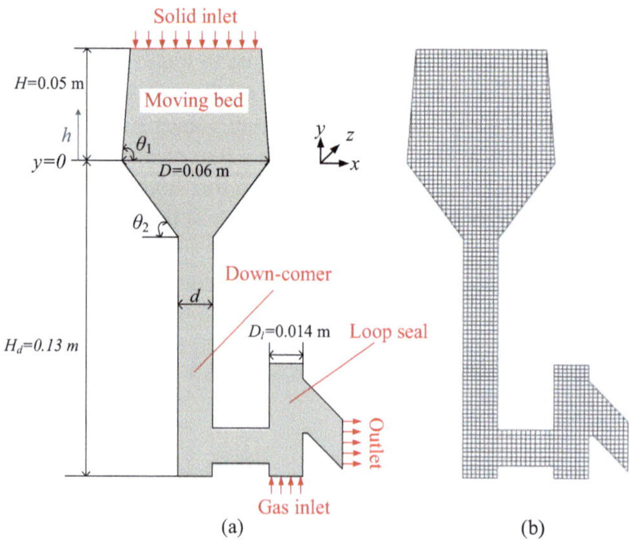

Fig. 6.10 Geometry and mesh of the moving bed air reactor [14]

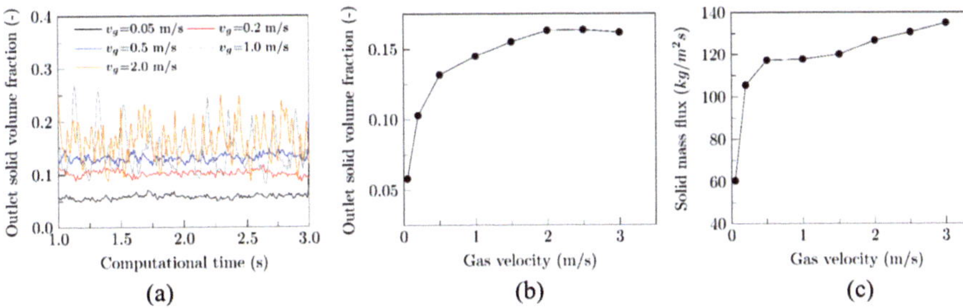

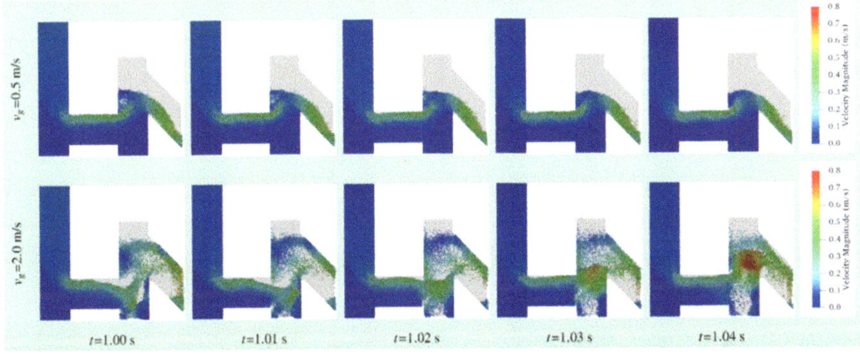

Fig. 6.11 (**a**) Instantaneous outlet solid volume fraction with time under different gas velocities, (**b**) average outlet solid volume fraction under different gas velocities, and (**c**) solid mass flux under different gas velocities [14]

Fig. 6.12 Instantaneous solid distributions under different gas velocities in the loop seal [14]

Figure 6.11 displays the instantaneous and time-averaged solid volume fraction at the outlet under different gas velocities. The outlet solid flows are stable between 1 and 3 s under involved conditions. When the gas velocity is 0.05, 0.2, and 0.5 m/s, the solid flow fluctuations are small, and the outlet solid volume fraction increases with increasing gas velocity in the loop seal. When the gas velocity increases to 1.0 and 2.0 m/s, intense periodical fluctuations are observed. Figure 6.11b shows that when the gas velocity is larger than 2.0 m/s, the average solid volume fraction at the outlet decreases slightly, which could be caused by the increasing solid velocity. Figure 6.11c displays the solid mass flux under different gas velocities. The solid mass flux first grows rapidly and then increases slowly with increasing gas velocity in the loop seal.

Figure 6.12 displays the instantaneous solid distributions under different gas velocities in the loop seal. When the gas velocity in the loop seal is 0.5 m/s, particles move smoothly to the outlet. When the gas velocity is 2.0 m/s, there are some intense solid clusters in the vertical pipe, which cause the periodical oscillations of the solid volume fraction at the outlet.

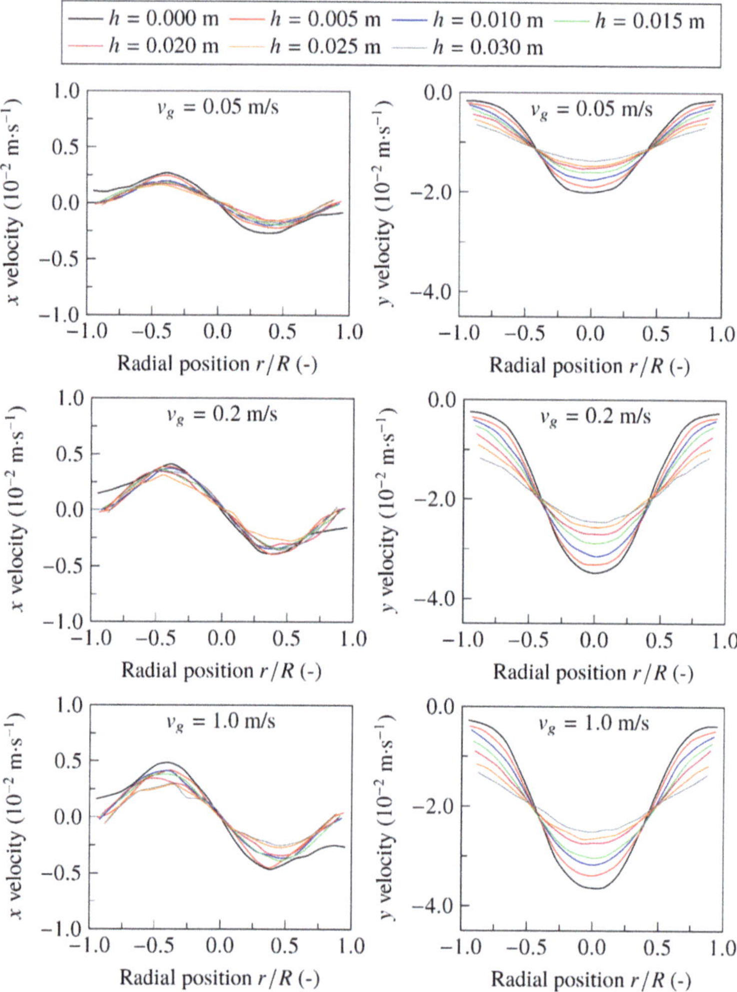

Fig. 6.13 Effects of the loop seal gas velocity on the time-averaged solid x and y velocity distributions [14]

Figure 6.13 displays effects of the loop seal gas velocity on the solid x and y velocity distributions. The y velocity is larger in the center of the reactor and smaller in the near-wall region. The central y velocity increases with decrease in the reactor height, while the near-wall y velocity decreases with decreasing bed height. The increase of the central y velocity is due to the effects of the wedge which is connected with the reactor at the bottom. To maintain the same solid flux in different cross-sections of the reactor, the near-wall y velocity decreases accordingly with decreasing reactor height. The y velocity curve is flatter under a smaller gas velocity. With increase in the gas velocity, the y velocity increases due to increasing solid

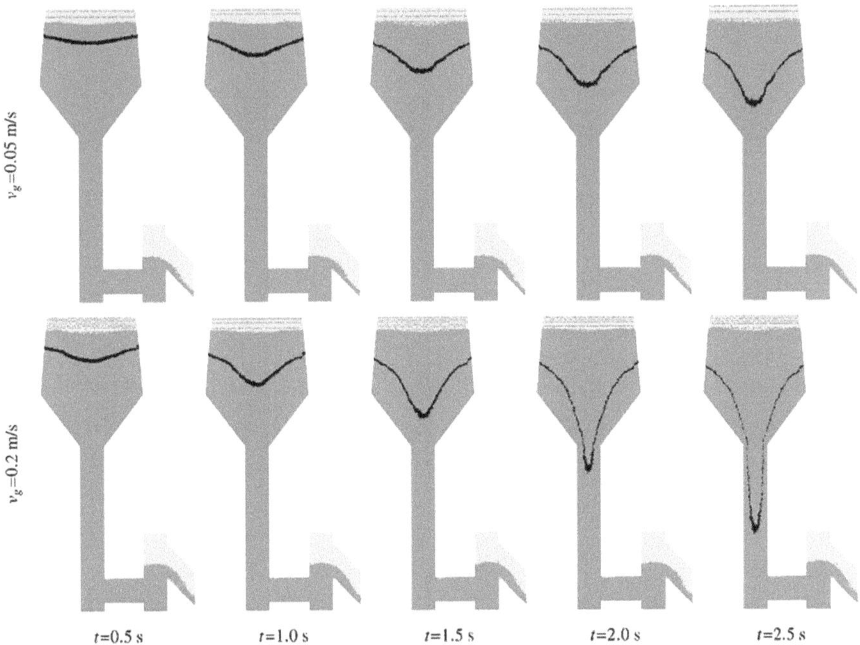

Fig. 6.14 Effects of the loop seal gas velocity on the instantaneous tracer particle distributions between 0.5 and 2.5s [14]

mass flux. From the x velocity distributions, it can be observed that almost all the particles move toward the center and the x velocity curve has two peaks. The x velocity increases with increasing gas velocity, which is related to the increasing solid mass flux.

Figure 6.14 displays the effects of the loop seal gas velocity on the instantaneous tracer particle distributions between 0.5 and 2.5 s. It is obvious that the particles move more quickly under the gas velocity of 0.2 m/s due to larger solid mass flux.

Figure 6.15a shows the accumulative solids' residence time distribution (RTD) under different loop seal gas velocities. Among the three cases, the tracer particles first leave the reactor under the gas velocity of 0.05 m/s. The difference between the curves of 0.05 and 0.2 m/s is more obvious than that between the curves of 0.2 and 1.0 m/s. Figure 6.15b presents that the average residence time of particles decreases with increase in the gas velocity. Figure 6.15c displays that the coefficient of variance first increases and then decreases with increasing gas velocity.

This chapter has provided examples of CFD-DEM simulations of a model fluidized bed reactor using cold flow and reacting flow gas injections and pseudo-coal injection as well as simulation of dense granular system in a model moving bed reactor. This approach can be applied to other fluidized bed or moving bed reactors.

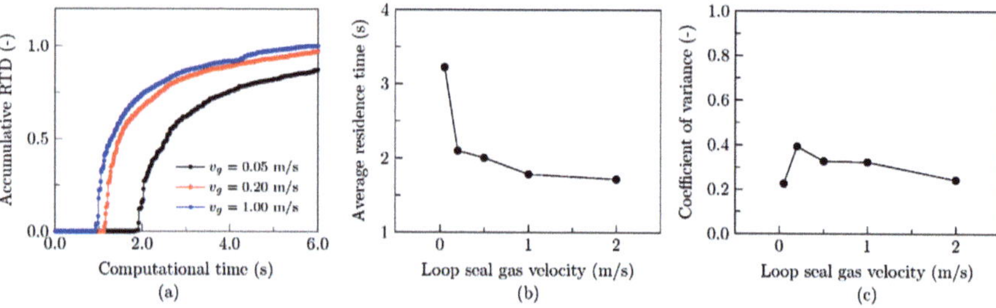

Fig. 6.15 (**a**) Accumulative solids' RTD under different loop seal gas velocities, (**b**) effects of the loop seal gas velocity on the average solids' residence time, and (**c**) effects of the loop seal gas velocity on the coefficient of variance [14]

References

1. J.M. Link, *Development and Validation of a Discrete Particle Model of a Spout-Fluid Bed Granulator* (University of Twente, Enschede, 1975)
2. V.S. Sutkar, N.G. Deen, B. Mohan, V. Salikov, S. Antonyuk, S. Heinrich, J.A.M. Kuipers, Numerical investigations of a pseudo-2D spout fluidized bed with draft plates using a scaled discrete particle model. Chem. Eng. Sci. **104**, 790–807 (2013)
3. F. Alobaid, J. Ströhle, B. Epple, Extended CFD/DEM model for the simulation of circulating fluidized bed. Adv. Powder Technol. **24**(1), 403–415 (2013)
4. Z. Zhang, R. Agarwal, Transient simulations of spouted fluidized bed for coal-direct chemical looping combustion, in *ASME Power Conference*, vol. 46087, (American Society of Mechanical Engineers, 2014), p. V001T01A018
5. S.R. Son, S.D. Kim, Chemical-looping combustion with NiO and Fe2O3 in a thermobalance and circulating fluidized bed reactor with double loops. Ind. Eng. Chem. Res. **45**(8), 2689–2696 (2006)
6. M.M. Hossain, H.I. de Lasa, Chemical-looping combustion (CLC) for inherent CO_2 separations—A review. Chem. Eng. Sci. **63**(18), 4433–4451 (2008)
7. M. Johansson, T. Mattisson, A. Lyngfelt, Investigation of Fe2O3 with MgAl2O4 for chemical-looping combustion. Ind. Eng. Chem. Res. **43**(22), 6978–6987 (2004)
8. D. Geldart, Types of gas fluidization. Powder Technol. **7**(5), 285–292 (1973)
9. D. Merrick, Mathematical models of the thermal decomposition of coal: 1. The evolution of volatile matter. Fuel **62**(5), 534–539 (1983)
10. D. Bradley, M. Lawes, H.-Y. Park, N. Usta, Modeling of laminar pulverized coal flames with speciated devolatilization and comparisons with experiments. Combust. Flame **144**(1–2), 190–204 (2006)
11. K. Mahalatkar, J. Kuhlman, E.D. Huckaby, T. O'Brien, CFD simulation of a chemical-looping fuel reactor utilizing solid fuel. Chem. Eng. Sci. **66**(16), 3617–3627 (2011)
12. H. Leion, T. Mattisson, A. Lyngfelt, Solid fuels in chemical-looping combustion. Int. J. Greenh. Gas Control **2**(2), 180–193 (2008)
13. Y. Shao, X. Wang, B. Jin, Y. Zhang, Z. Kong, X. Wang, Z. Jin, Carrying capacity and gas flow path mechanism of a novel multistage air reactor for chemical looping combustion. Energy Fuels **32**(12), 12665–12678 (2018)
14. Y. Shao, R.K. Agarwal, J. Li, X. Wang, B. Jin, Computational fluid dynamics-discrete element model simulation of flow characteristics and solids' residence time distribution in a moving bed air reactor for chemical looping combustion. Ind. Eng. Chem. Res. **59**(40), 18180–18192 (2020)

Chapter 7
CFD Simulations of a Single Reactor for CLC

7.1 Bubbling Fluidized Bed Fuel Reactor

To understand the bubbling fluidized bed behavior, Wang et al. [1] simulated a 2D bubbling fluidized bed with a width of 25 mm and a height of 80 mm as shown in Fig. 7.1 by employing the two-fluid model. NiO supported with bentonite was used as the oxygen carrier (OC) to increase the reactivity and durability of oxides. The gaseous fuel CH_4 was used; the gas-solid reaction between the fuel and oxygen carrier can be expressed as $CH_4(g) + 4NiO(s) \Rightarrow CO_2(g) + 2H_2O(g) + 4Ni(s)$. The OC particle has a diameter of 120 μm and a density of 3598 kg/m^3. OC particles consisted of 57.8% NiO and 42.2% bentonite by mass fraction. Initially, particles were packed in the reactor with a height of 25 cm and solid volume fraction of 0.58, while the rest of the space was filled with N_2. The gas was injected into the reactor from the bottom with superficial velocity of 10 cm/s. The top of the reactor was set as the Neumann boundary condition which didn't allow the particles to flow out the reactor. The initial temperature of the reactor was 950 K. The restitution coefficient between the particles and wall restitution coefficient were set as 0.97 and 0.95, respectively. From 0 to 2 s, the inlet gas was totally N_2 to avoid numerical divergence; after 2 s, the inlet gas composition was changed to 90% CH_4 and 10% N_2 by mass fraction. At the wall, the no-slip boundary condition was specified for the gas flow, which means that both the tangential and normal components of gas velocity were set to zero. A constant time step of 10^{-5} s was used in the numerical simulation.

Figure 7.2 displays the instantaneous solid distributions in a bubbling fluidized bed fuel reactor. Small bubbles formed near the distributor gradually move to the center of the reactor during the upward movement. Meanwhile, bubbles grow in size by merging with adjacent small bubbles on the way up. Low-pressure zones are formed in the wake of large bubbles, and thus the leading bubbles absorb the trailing small bubbles, and particles are drawn into the wake. Bubbles move upward continuously and finally break at the interface between the dense phase and free

R. K. Agarwal, Y. Shao, *Modeling and Simulation of Fluidized Bed Reactors for Chemical Looping Combustion*, https://doi.org/10.1007/978-3-031-11335-2_7

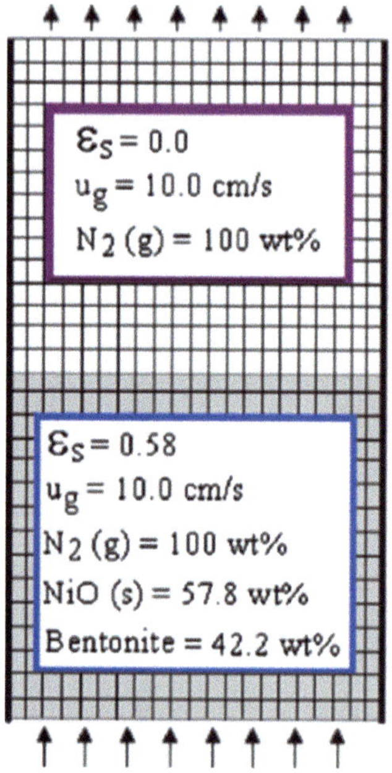

Fig. 7.1 Initial conditions and parameters used in the simulations [1]

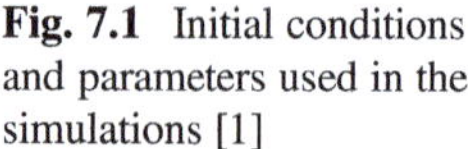

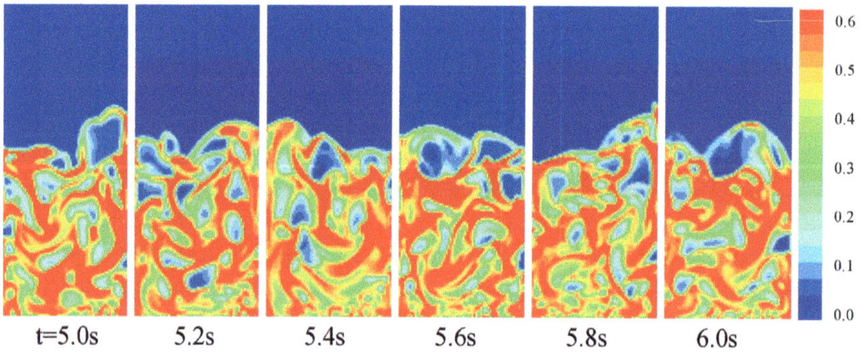

Fig. 7.2 Instantaneous solid distribution in a BFB fuel reactor [1]

board region. Driven by gravity, the particles moved to the freeboard and finally drop back to the bed. The process of bubble formation and rise and burst in the BFB fuel reactor were also reported by Jung et al. [2], Deng et al. [3], Wang et al. [4], Harichandan et al. [5], and others. The existence of bubbles especially those with large size and fast velocity inevitably affects the reaction of the gaseous fuel. The combustible gas may flow through the bubbles without contacting and reacting with the oxygen carriers, which leads to decrease in the fuel conversion efficiency. In a bubbling fluidized bed, the bubble size and the degree of mixing of gas and solid phases are affected by the bed height, bed width, free-board height, particle sizes, etc.

7.2 Spouted Fluidized Bed Fuel Reactor

Different from a BFB fuel reactor, a SFB fuel reactor creates a stronger mixing of gas and solid particles with the introduction of a high-velocity jet in order to avoid the ash agglomeration with the oxygen carrier during the reaction process [6]. A quasi-3D laboratory-scale experimental SFB fuel reactor was built at the Darmstadt University of Technology (TU-Darmstadt). Zhang et al. [6] simulated the gas-solid flow dynamics in the SFB fuel reactor using the Eulerian-Lagrangian approach. The experimental and numerical models are displayed in Fig. 7.3.

As shown in Fig. 7.4, with the high-speed gas injected into the reactor, a spout region is formed in the center which is surrounded by an annular dense region. The particles in the spout region are carried by the high-speed gas to the top of the bed and get blown above the bed. Without sufficient momentum from the gas, particles flown to the top are then driven by gravity to flow downward to the surface of the annular dense region and move with other particles. At the bottom of the bed, particles in the dense phase are continuously fed in the spout region. Thus, a solid circulation between the central spout region and annular dense region is formed.

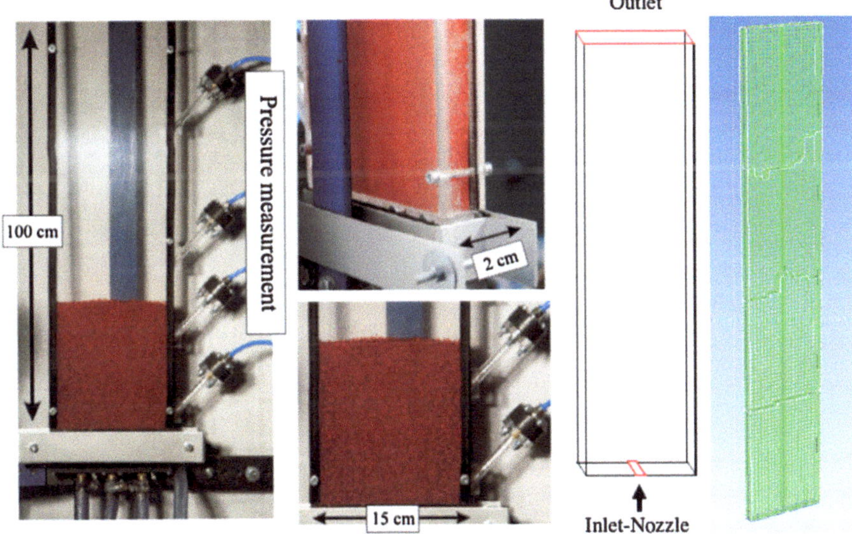

Fig. 7.3 Test rig of the spouted fluidized bed apparatus at Darmstadt University of Technology (left) and CFD model (right) [6]

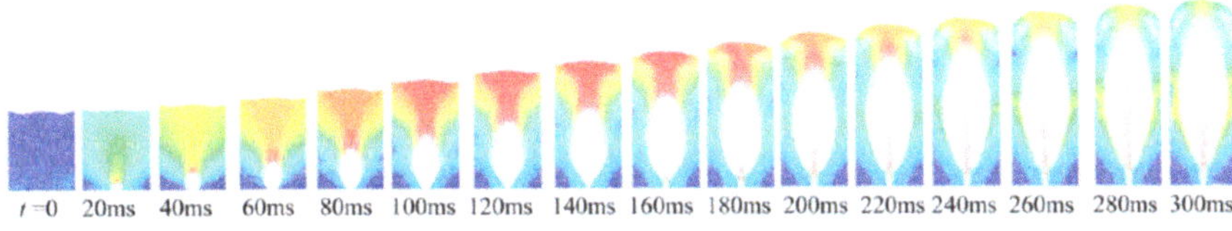

Fig. 7.4 Instantaneous solid distribution in a SFB fuel reactor [6]

7.3   Circulating Fluidized Bed Riser Fuel Reactor

To enhance the conversion of solid fuel and lengthen the lifetime of oxygen carrier, a fuel reactor including a gasifier reactor and a reduction reactor was proposed by Wang et al. [7] as shown in Fig. 7.5. The bottom gasifier reactor is a bubbling fluidized bed, while the top reduction reactor is operated under the fast fluidization regime. In the gasifier reactor, the solid fuel such as coal is gasified within the inertia bed material, namely, the silica sand. The oxygen carrier is fed into the reduction reactor from the bottom feeding section to react with the gasification product from the gasifier reactor. The relatively low velocity in the gasifier reactor is beneficial for providing long residence time for the solid fuel, and the strong solid mixing and internal circulation can also promote the solid fuel gasification process. The

Fig. 7.5 Sketch of the fuel reactor with a gasifier and a reduction reactor [7]

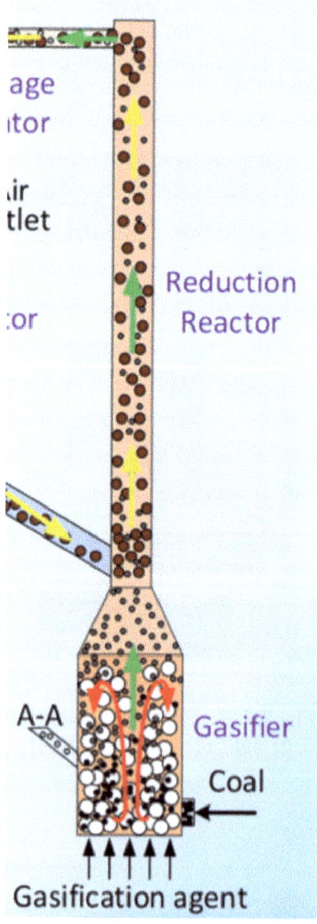

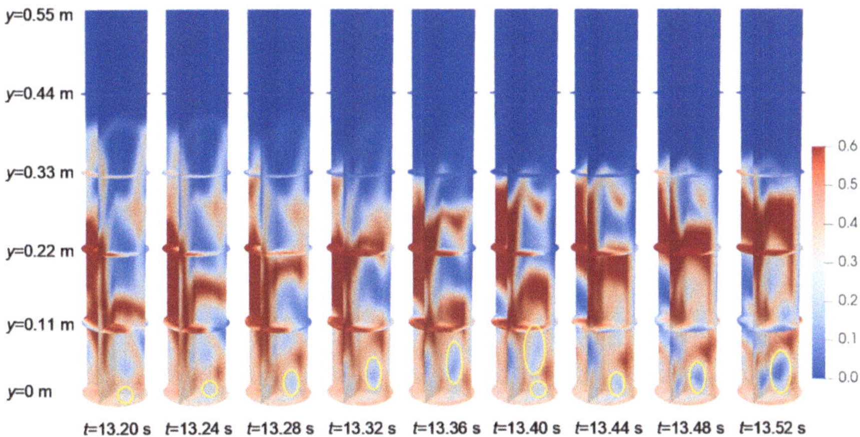

Fig. 7.6 Instantaneous solid volume fraction in the gasifier reactor [7]

turbulent gas-solid flow in the reduction reactor offers sufficient gas-solid contact and further enhances the reduction of the gasification products by the oxygen carrier. To investigate gas-solid flow characteristics in the designed fuel reactor, Wang et al. [8] built a 3D numerical model and employed the Eulerian-Eulerian method for numerical simulation. In the fuel reactor, the inner diameter and height of the gasifier reactor were 0.10 m and 0.55 m, respectively; the reduction reactor was a riser with an inner diameter of 0.06 m and a height of 2.05 m; a converging pipe with a height of 0.125 m was used to connect the gasifier reactor and reduction reactor.

Figure 7.6 shows the instantaneous solid volume fraction obtained from the simulation in the bottom gasifier reactor. Similar to the flow behaviors of bubbling fluidized bed described in Sect. 7.1, the typical process of bubble generation, their merging, rising, and breakup, is clearly observed in the gasifier reactor. The gasifier reactor operates in the bubbling fluidization state as expected, and almost no sand particles are carried by the gas to the freeboard region. Figure 7.7 displays the distribution of the gas phase in the reduction reactor during the quasi-stable process. The solids behavior in the riser can be categorized into three sections which are the bottom high-density section, middle transition section, and upper dilute section. Because of the high superficial gas velocity, bubbles are not formed in the riser which can therefore avoid the escape of combustible gas through bubbles and further enhance the reaction efficiency. Compared to the bubbling fluidized bed and spouted fluidized bed, the intense flow regime in a circulating fluidized bed riser can provide sufficient gas-solid contact over the entire or majority of the reactor height, which is beneficial for improving the reaction efficiency.

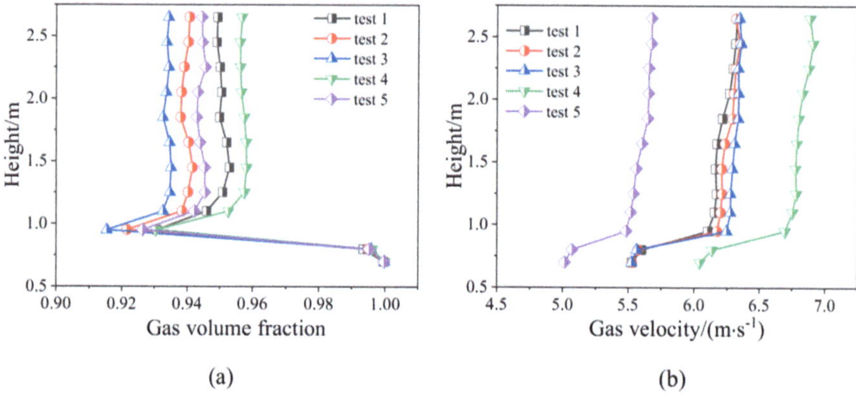

Fig. 7.7 Operational state of the gas phase in reduction reactor during quasi-stable process: (**a**) volume fraction; (**b**) velocity [7]

References

1. W. Shuai, Y. Yunchao, L. Huilin, W. Jiaxing, X. Pengfei, L. Guodong, Hydrodynamic simulation of fuel-reactor in chemical looping combustion process. Chem. Eng. Res. Des. **89**(9), 1501–1510 (2011)
2. J. Jung, I.K. Gamwo, Multiphase CFD-based models for chemical looping combustion process: Fuel reactor modeling. Powder Technol. **183**(3), 401–409 (2008)
3. Z. Deng, R. Xiao, B. Jin, Q. Song, Numerical simulation of chemical looping combustion process with $CaSO_4$ oxygen carrier. Int. J. Greenh. Gas Control **3**(4), 368–375 (2009)
4. X. Wang, B. Jin, Y. Zhang, W. Zhong, S. Yin, Multiphase computational fluid dynamics (CFD) modeling of chemical looping combustion using a CuO/Al_2O_3 oxygen carrier: Effect of operating conditions on coal gas combustion. Energy Fuels **25**(8), 3815–3824 (2011)
5. A.B. Harichandan, T. Shamim, CFD analysis of bubble hydrodynamics in a fuel reactor for a hydrogen-fueled chemical looping combustion system. Energy Convers. Manag. **86**, 1010–1022 (2014)
6. Z. Zhang, L. Zhou, R. Agarwal, Transient simulations of spouted fluidized bed for coal-direct chemical looping combustion. Energy Fuels **28**(2), 1548–1560 (2014)
7. X. Wang, X. Wang, S. Zhang, Z. Kong, Z. Jin, Y. Shao, B. Jin, Test operation of a separated-gasification chemical looping combustion system for coal. Energy Fuel **32**(11), 11411–11420 (2018)
8. X. Wang, Y. Shao, B. Jin, Spatiotemporal statistical characteristics of multiphase flow behaviors in fuel reactor for separated-gasification chemical looping combustion of solid fuel. Chem. Eng. J. **2021**(412), 128575 (2020, October)

Chapter 8
Full-Loop Simulations of Chemical Looping Systems

8.1 Single-Loop Circulating Fluidized Bed Chemical Looping System

8.1.1 Chemical Looping System Utilizing a Moving Bed Air Reactor

Jin et al. [1] proposed and built a single-loop CLC system which consisted of a circulating fluidized bed (CFB) riser as the fuel reactor (FR) and a moving bed as the air reactor (AR) as shown in Fig. 8.1a. The system has the advantages of simple structure, stable operational performance, and flexibility in adjustment. The FR is operated under the turbulent fast fluidization state, which can ensure sufficient gas-solid contact in most regions of the reactor. The AR is located in the middle of the downcomer and is designed to operate under the moving bed state, which has the advantages of low-pressure drop and favorable gas-solid contact. Only one driving source is required in the system which is originated from the circulating fluidized bed riser, and the adjustment of AR operation has little impact on the quasi-stable solid circulation in the whole system. This enhances the operational flexibility of the system. For this system, Shao et al. [2] built a 3D numerical model and employed the Eulerian-Eulerian approach to study it. Simulation results show that the quasi-steady solid circulation was successfully achieved along the loop of "FR-separator-down comer-AR-down comer-loop seal-FR," as shown in Fig. 8.1b. The bed height in the upper downcomer fluctuated steadily, which is regarded as the sign of the realization of the quasi-stable solid circulation. The key point of establishing a quasi-stable solid circulation is to build reasonable pressure balance in the whole system. The largest pressure should occur in the loop seal so that the particles can be continuously transported from the downcomer to the FR via the loop seal. In addition, the high pressure in the loop seal is able to prevent the gas from leaking

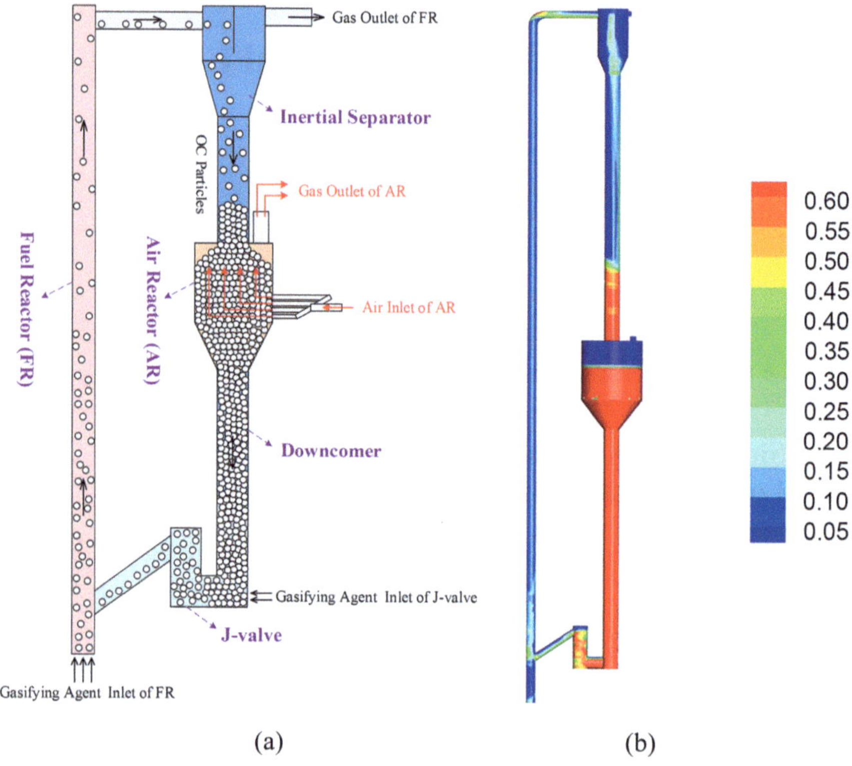

Fig. 8.1 Chemical looing combustion system based on moving bed air reactor [2]: (**a**) schematic of the system, (**b**) simulated solid distributions in the system

from the FR to the downcomer. Otherwise, large amount of gas flowing to the loop seal would cause the particles in the FR not to be carried upward to the top of FR by sufficient momentum from the gas.

To further improve the carrying capacity of gas flow in the AR, a multi-stage moving bed AR was proposed. Compared with the single gas outlet located at the top of the moving bed shown in Fig. 8.1, the gas outlets in the multi-stage AR are groove channels which are inserted to the moving bed as shown in Fig. 8.2, which enables multi-stage gas-solid contact and reaction and greatly improves the amount of gas allowed to be fed into the moving bed. The gas-solid distribution uniformity in the novel AR and complex gas flow paths were investigated [3]. Results showed that the solid velocity uniformity index is smaller than 0.7 due to effects of internals, while solid concentration uniformity index is higher with values larger than 0.98. Besides, the gas flow paths in the system can be optimized via the regulation of pressure difference. The optimized gas flow paths are shown in Fig. 8.2, with minimal gas leakage between the separator and multi-stage AR.

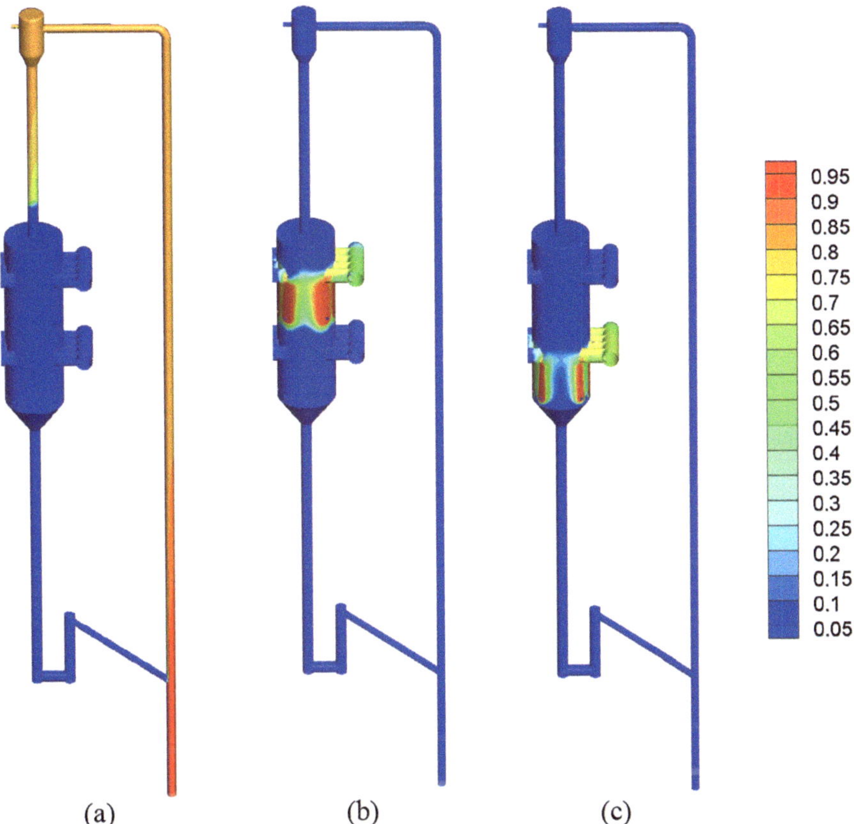

Fig. 8.2 Gas mole fraction distributions in the whole system after the regulation of gas flow paths [3]: (**a**) FR gas; (**b**) first-stage AR inlet gas; (**c**) second-stage AR inlet gas

8.1.2   *Chemical Looping System Based on a Bubbling Fluidized Bed Fuel Reactor*

Another type of CLC system described here is based on the circulating fluidized bed. An extra bubbling fluidized bed located after the cyclone is used as the FR, while the riser is used as the AR which is fluidized above the terminal velocity [4]. The principal layout of the system is illustrated in Fig. 8.3.

For this system, Wang et al. [5] established a 2D two-fluid model to simulate the flow behavior and gas leakage in a single-CFB CLC system. Figure 8.4a shows the particle distribution predicted by the simulation. Particles are carried by the gas to move upward in the high-velocity AR in which the core-annular structure is formed. Particles are then dragged to the cyclone and flow down along dipleg to the low-velocity FR. Particles further flow from the FR to the low-velocity pot-seal and then to the AR by gravity and waft of the gas. The particle circulation is achieved

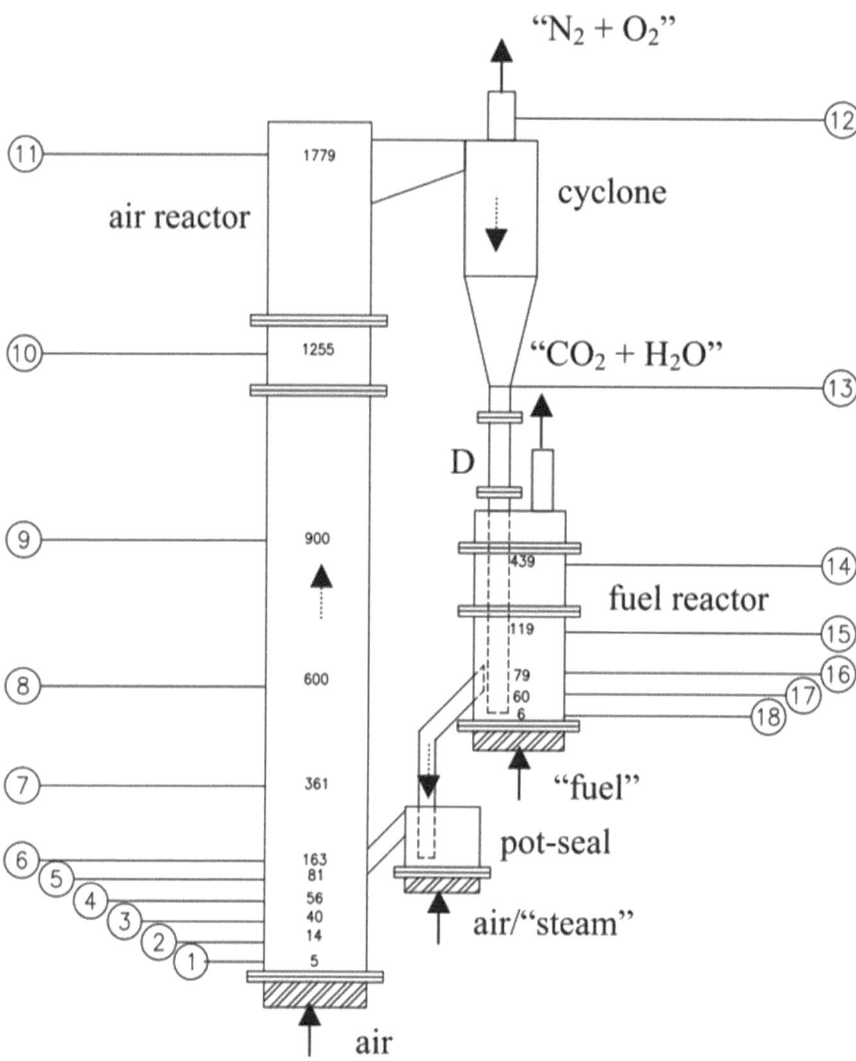

Fig. 8.3 Principal layout of the CLC system based on bubbling fluidized bed [4]

by overflow in the FR and pot-seal, inevitably causing gas leakage between different compartments. In the 2D simulation, it was demonstrated that the solid inventory, FR inlet gas velocity, AR inlet gas velocity, and pot-seal inlet gas velocity all have influence on the gas leakage ratio. The influence of the bed walls and geometry on nonsymmetric flow cannot be considered in a 2D model. Guan et al. [6] developed a 3D model for the same experimental apparatus as studied by Wang et al. [5] and the results for the solid distributions obtained by them are shown in Fig. 8.4b. In the 3D simulation, bubbles are more obvious in the BFB FR, and the function of the cyclone is modeled. Furthermore, Ahmed et al. [7] performed a reactive flow simulation to

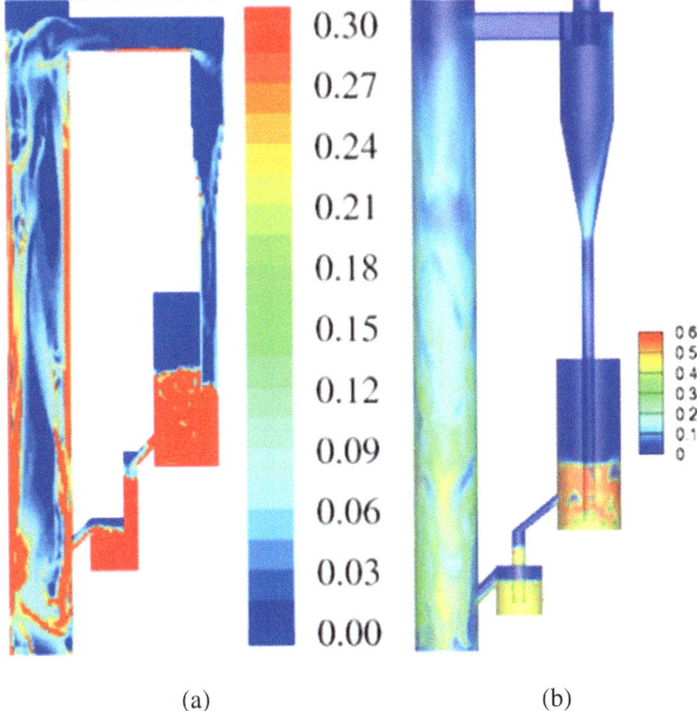

Fig. 8.4 Instantaneous particle distribution in a single-CFB CLC system: (**a**) 2D simulation [5], (**b**) 3D simulation [6]

study the system performance in the presence of reactions in a similar single-loop CLC system with NiO as the oxygen carrier and CH_4 as the fuel; the gas component distributions are shown in Fig. 8.5. The CH_4 concentration decreases along the bed height in the FR, and the concentrations of intermediate products such as H_2 and CO show similar trends. The unconverted reactant is caused by the bubbles which provide shortcuts for gas flow. The bubble movements also lead to the oscillations of CH_4, CO_2, and H_2O at the FR outlet. The AR is composed of a bottom BFB and a top riser. The O_2 concentration decreases along the bed height, which demonstrates more particles are regenerated during the oxidation process. The solid mass flux oscillates greatly at the AR outlet due to the formation and breaking of clusters.

The National Energy Technology Laboratory (NETL) in the USA also designed a CLC system based on the bubbling fluidized bed [8], as shown in Fig. 8.6a. The FR and AR both operate under the bubbling fluidized state. Parker et al. [9] investigated the coal-fired iG-CLC process in the system using the computational particle fluid dynamics (CPFD) method. Figure 8.6b, c shows the solid distribution colored by temperature and instantaneous temperature distributions in the two reactors. The temperature in the AR is higher than that in the FR as a result of exothermic oxidation reaction and endothermic reduction and gasification reactions.

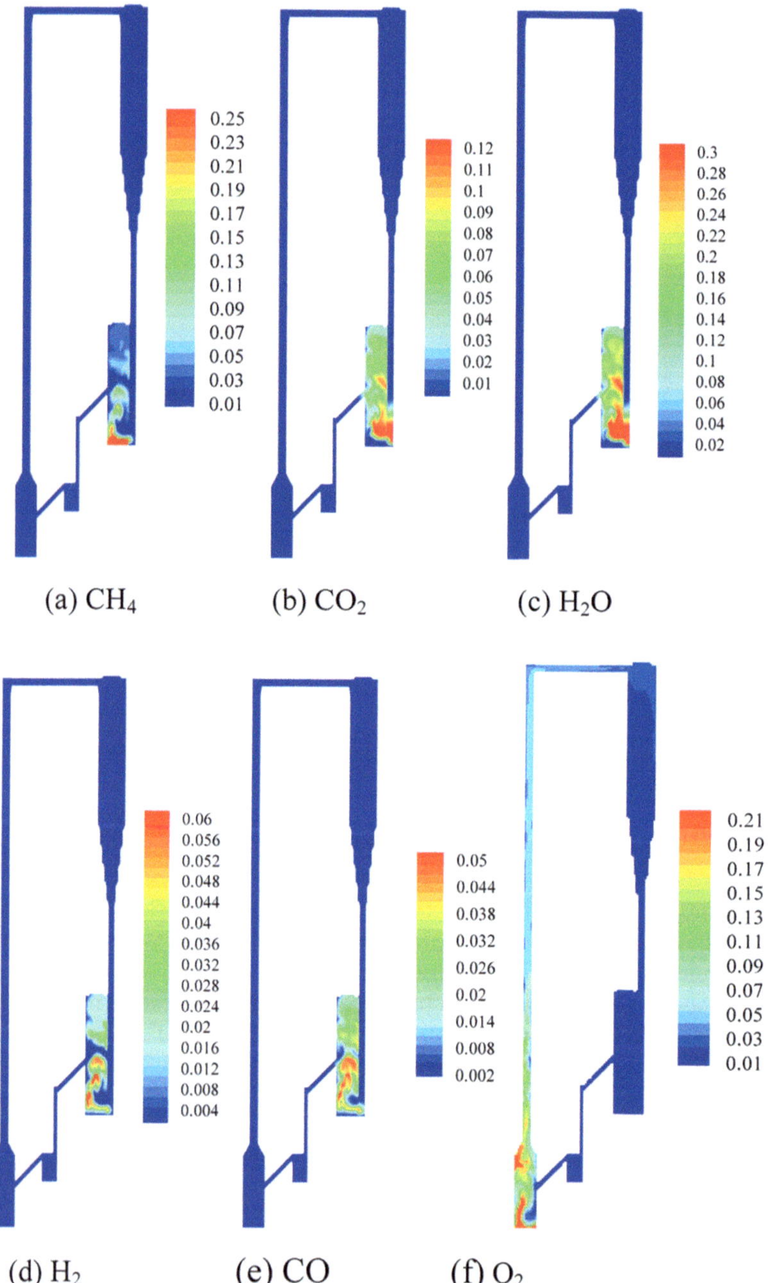

Fig. 8.5 Instantaneous gas components distribution in a single-CFB CLC system [7]: (**a**) CH$_4$, (**b**) CO$_2$, (**c**) H$_2$O, (**d**) H$_2$, (**e**) CO, (**f**) O$_2$

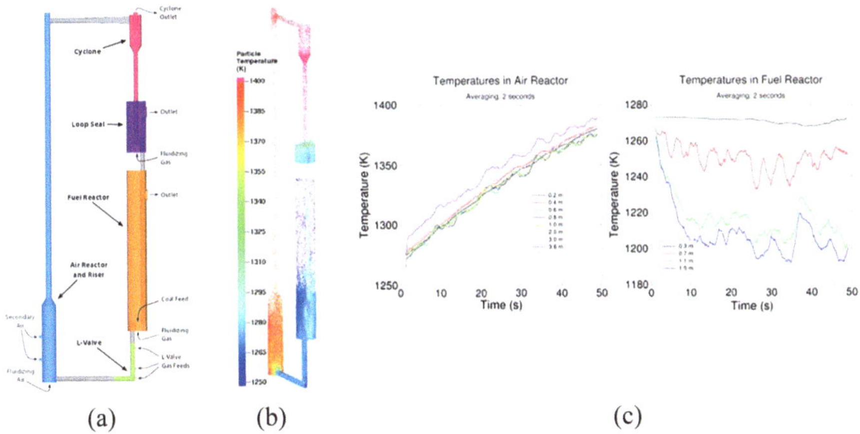

Fig. 8.6 (**a**) Principal layout of the CLC system based on bubbling fluidized bed [9], (**b**) instantaneous solid distribution colored by temperature [9], and (**c**) temperature distributions in two reactors [9]

Temperature increase in the AR and temperature decrease in the FR can be observed since the start of the reaction. Solid circulation rate is an important parameter to reveal the continuity and stability of solid running between the FR and AR.

Another similar single-loop CLC system shown in Fig. 8.7 was simulated by Peng et al. [10] employing the dense discrete phase model (DDPM) with discrete element method (DEM). Two fluidized loop seals were used in the experimental setup with the upper loop seal located before the fuel reactor to prevent leakage of air to the fuel reactor, and the lower loop seal was located after the fuel reactor to prevent any leakage of fuel to the air reactor. In the cold-flow mode of simulation [10], the total number of parcels of solid particles was 241,685. The solid circulation rate increased with increasing solid inventory and increasing inlet gas flow rates of the FR, AR, and loop seal. Furthermore, the reaction of CH_4 with NiO was incorporated in the 3D model [11]. The solid distributions colored by solid mass conversion, solid temperature, and gas temperature are shown in Fig. 8.8b–d. The solid mass conversion rate x which is defined as $x = (m - m_{red})/(m_{ox} - m_{red})$ is a measure of how much oxygen atoms have been transferred from the AR to the FR by oxygen carriers. Due to the oxidation reaction in the AR and reduction reaction in the FR, the solid mass conversion rates in the AR, cyclone, and upper loop seal were obviously larger than those in the FR and lower loop seal. In different parts of the CLC system, the distributions of mass conversion rate are not uniform, which are caused by the movements of the bubbles and clusters. The gas temperature distribution is in line with the solid temperature distribution because of the heat transfer between the solid phase and gas phase.

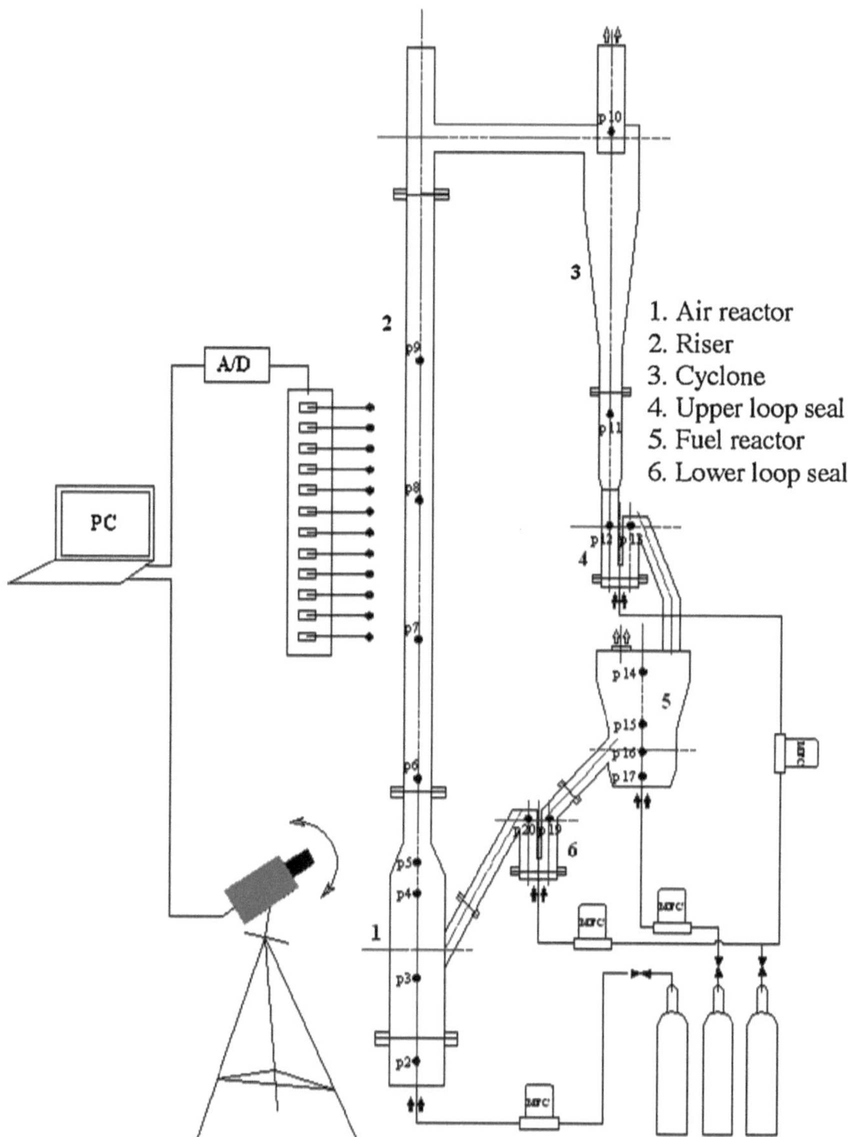

Fig. 8.7 Schematic drawing of the experimental setup [10]

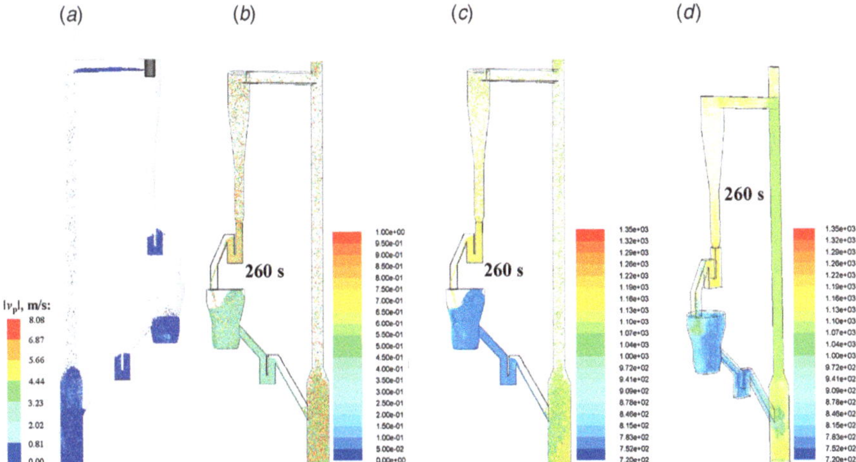

Fig. 8.8 Flow and reaction performance in a single-CFB CLC system: (**a**) particle distributions obtained from cold-flow mode simulation [10], (**b**) solid distribution colored by solid mass conversion rate [11], (**c**) solid distribution colored by solid temperature [11], and (**d**) solid distribution colored by gas temperature [11]

8.2   Dual-Loop Circulating Fluidized Bed Chemical Looping System

The dual-CFB CLC system has a more complex structure than the single-CFB CLC system. The solid circulation between the two CFBs, optimization of operation performance, effects of clusters, and reaction process are all investigated in different numerical simulations.

8.2.1   System from Huazhong University of Science and Technology

Ma et al. [12] designed and constructed a 50 kW$_{th}$ CLC unit based on dual circulating fluidized bed shown in Fig. 8.9. The CLC system mainly consists of a bubbling fluidized bed fuel reactor, a turbulent fluidizing air reactor, risers, loop seals, a carbon stripper, cyclones, and downcomers. The reaction performance of the system, the combustion efficiency, CO_2 yield, and carbon capture efficiency were investigated in the experiment. However, some operation characteristics were still lacking including the solid circulation rate, solid residence time, reaction details, etc.

Chen et al. [13] adopted the CPFD approach to build a 3D reactive model for the dual-CFB CLC system in Barracuda software to investigate the gas-solid reactive

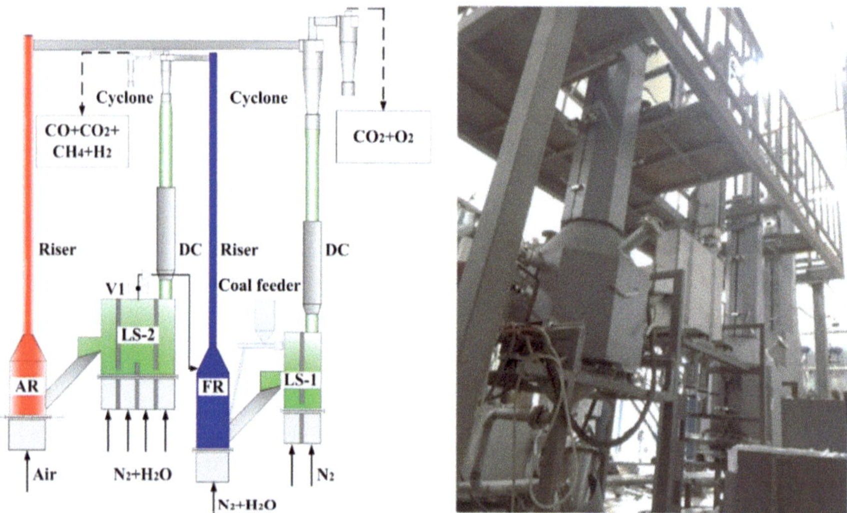

Fig. 8.9 Experimental setup of Huazhong University of Science and Technology [12]

Table 8.1 Proximate and ultimate analysis of the Shenhua bituminous coal [13]

Proximate analysis (ad wt.%)			
Fixed carbon	48.46	Volatile	15.54
Moisture	1.66	Ash	34.34
Ultimate analysis (ad wt.%)			
C	55.26	H	2.12
O	5.39	N	0.79
S	0.44		

flow and optimize the reactor structure. Hematite was used as the oxygen carrier (OC), and Shenhua bituminous coal was used as the solid fuel, which are the same material as those used in the experiment. The OC has a diameter range of 150–350 μm and density of 3650 kg/m³. The coal size ranges between 100 and 300 μm, and the proximate and ultimate analyses are shown in Table 8.1. The total number of parcels of particles is about 2,139,908. Cartesian mesh is used with the total grid number of 476,000.

The particle distribution is shown in Fig. 8.10a. The residence time of char is too short to be totally converted, and the one-point coal feeding mode of the FR cannot ensure favorable mixing of the oxygen carrier and combustible gases. Increasing the FR bed height and adopting a two-point coal feeding mode are two effective methods to improve the combustion efficiency. However, there were still half of the char particles escaping to the AR in the simulation. To reduce the amount of particles flowing to the AR, different types of gasification carbon strippers (GCS) and carbon strippers (CS) can be placed at the outlet of the FR. Chen et al. [14] placed three types of GCS and CS within the loop seal at the FR outlet and tested

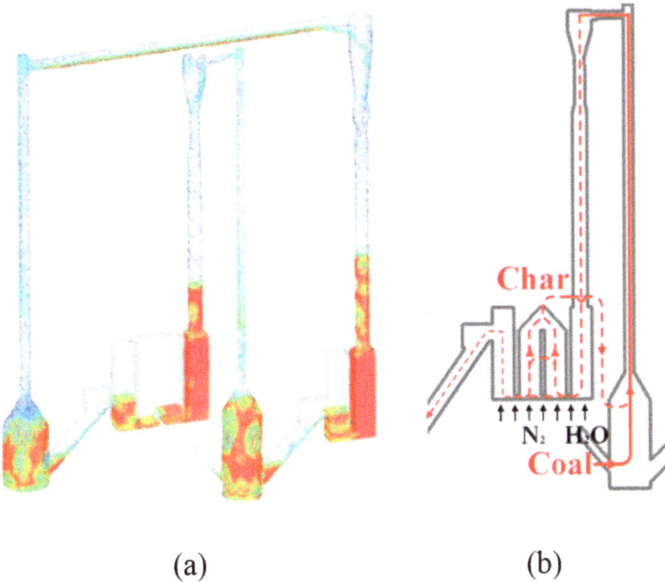

Fig. 8.10 (**a**) Solid distribution in a dual-CFB CLC system [13] and (**b**) optimal strategy to improve the carbon conversion [14]

their effects on improving the carbon conversion using the 3D reactive model. The optimal strategy shown in Fig. 8.10b was finally determined. The newly designed CS coupled within the loop seal can make a balance between the operational cost and performance, which helps the CLC system achieve lower char slip, higher carbon conversion efficiency, and relatively lower operational cost.

For a similar coal-fired dual-CFB CLC system, Su et al. [15] built a 2D two-fluid model to study the gas leakage, flow patterns, and gas component distributions. The operational performance in the FR was optimized by decreasing the FR coal feeding rate and increasing the FR temperature.

8.2.2 System from Vienna University of Technology

A 120 kW CLC system based on dual-CFBs was built at Vienna University of Technology [16, 17]. As shown in Fig. 8.11, the AR is designed as a fast fluidized bed with pneumatic transport of the particles. The oxygen carrier is oxidized with air and transported via a loop seal (LS) to the FR which is operated in turbulent regime. The introduced fuel in the FR is oxidized reducing the OC which is then transported back to the AR via a second LS.

The heterogeneous gas-solid flow with dense "cluster" phase and dilute "broth" phase is a significant feature in a CFB [18]. For a CLC system with two CFBs, it is essential to consider the effects of heterogeneity. A cluster structure-dependent

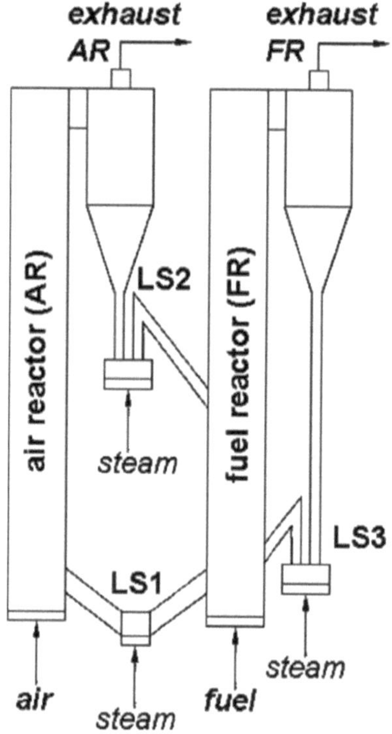

Fig. 8.11 Experimental setup of Vienna University of Technology [17]

(CSD) drag coefficient model can be incorporated to simulate the flow behavior in a dual-CFB CLC system, since the CSD model can consider the mesoscale heterogeneous structures involving clusters or strands [18]. Wang et al. [18] performed a 2D simulation incorporating the CSD model to simulate the gas-solid flow dynamics in the CLC system from Vienna University of Technology. Figure 8.12 illustrates the instantaneous solid distribution in a dual-CFB CLC system. The distributions of solid velocity, mass flux, and pressure were also investigated in detail. Further, the reaction between CH_4 and NiO was added to the 2D two-fluid model [19]. The gas temperature and gas species distributions in the whole system were predicted, but the effects of clusters on reactions and interphase heat transfer were not considered in the reactive model. To improve the computational accuracy, a multi-scale chemical reaction model coupling heat transfer was then adopted to fully reflect the multi-scale structure impacts [20], where the simulation results agree better with experimental data than the conventional model.

The solid circulation in a dual-CFB CLC system is more complex than that in a single rector simulation. For the dual CFB system from the Vienna University of Technology [16], Geng et al. [21] established a 3D cold-flow model to study the solid circulation rate distributions as shown in Fig. 8.13. The internal solid circulation rate and global solid circulation rate are FR solid flux and AR solid flux, respectively. The gas flow rates of FR and loop seal have a different influence on the two solid circulation rate and their fluctuations.

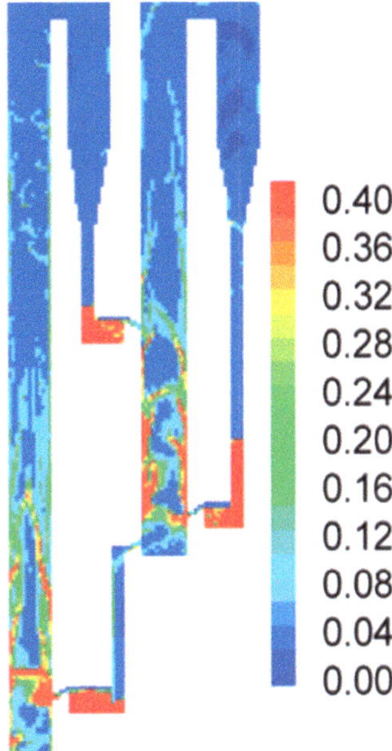

Fig. 8.12 Instantaneous solid distribution in a dual-CFB CLC system: 2D simulation by Wang et al. [18]

8.2.3   *System from the University of Utah*

A dual circulating interconnected fluidized bed chemical looping process unit with thermal power of 220 kW_{th} was developed at the University of Utah as shown in Fig. 8.14a. The FR and AR have bed heights of 6.0 and 5.7 m, respectively. The FR is a circulating fluidized bed which is fluidized by superheated steam and operated in the turbulent circulating regime. Fine char particles are carried by the gas to move upward in the FR and then separated from the flue gas via the cyclone. Then, fine char particles move downward to the dipleg by gravity and return back to the FR via the FRR loop seal. The coarse oxygen carriers are transported to the AR via the FR-to-AR loop seal. The loop seal is fluidized with a combination of air and steam, and the steam is used in the inlet close to the FR so as to prevent the N_2 from flowing to the FR; otherwise, the CO_2 purity in the FR outlet will be decreased due to the existence of the N_2. The AR is also a circulating fluidized bed which is fluidized by preheated gas with velocity of 5.0 m/s. Oxidized oxygen carriers move upward and flow through the cyclone, dipleg, and AR-to-FR loop seal to finally enter the FR. Reinking [22] developed a reactive 3D model for a dual-CFB CLOU system using Barracuda-VR™ software. The oxygen carrier is copper oxide and the solid fuel is coal. The total number of particles is 3.3×10^{10} with 8×10^5 parcels of particles. Figure 8.14b, c displays the solid distribution and char distribution. In the simulation, the quick reactions of volatiles occur in the FR and over 90% carbon capture efficiency is obtained, which demonstrates the high reaction efficiency of CLOU. However, the model validation is not carried out due to the lack of experimental data.

Fig. 8.13 Instantaneous solid distribution in a dual-CFB CLC system: 3D simulation by Geng et al. [21]

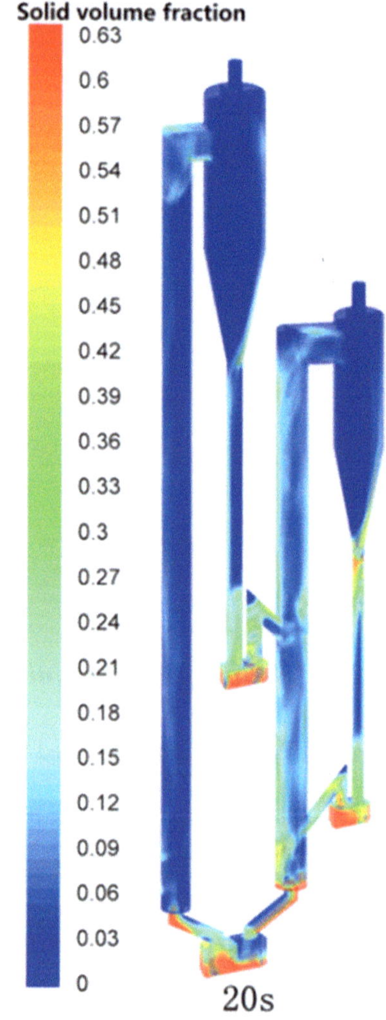

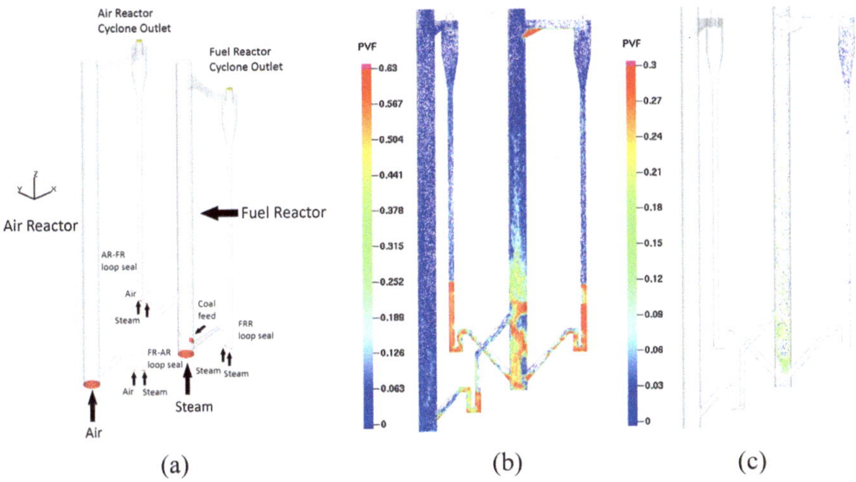

(a) (b) (c)

Fig. 8.14 (**a**) Experimental setup of University of Utah [22], (**b**) particle volume fraction [22], (**c**) particle volume fraction of char carbon [22]

References

1. X. Wang, B. Jin, H. Liu, W. Wang, X. Liu, Y. Zhang, Optimization of in situ gasification chemical looping combustion through experimental investigations with a cold experimental system. Ind. Eng. Chem. Res. **54**(21), 5749–5758 (2015)
2. Y. Shao, Y. Zhang, X. Wang, X. Wang, B. Jin, H. Liu, Three-dimensional full loop modeling and optimization of an in situ gasification chemical looping combustion system. Energy Fuels **31**(12), 13859–13870 (2017)
3. Y. Shao, R.K. Agarwal, X. Wang, B. Jin, Numerical simulation of a 3D full loop IG-CLC system including a two-stage counter-flow moving bed air reactor. Chem. Eng. Sci. **217**, 115502 (2020)
4. E. Johansson, A. Lyngfelt, T. Mattisson, F. Johnsson, Gas leakage measurements in a cold model of an interconnected fluidized bed for chemical-looping combustion. Powder Technol. **134**(3), 210–217 (2003)
5. W. Shuai, L. Guodong, L. Huilin, C. Juhui, H. Yurong, W. Jiaxing, Fluid dynamic simulation in a chemical looping combustion with two interconnected fluidized beds. Fuel Process. Technol. **92**(3), 385–393 (2011)
6. Y. Guan, J. Chang, K. Zhang, B. Wang, Q. Sun, Three-dimensional CFD simulation of hydrodynamics in an interconnected fluidized bed for chemical looping combustion. Powder Technol. **268**, 316–328 (2014)
7. A. Bougamra, L. Huilin, Modeling of chemical looping combustion of methane using a Ni-based oxygen carrier. Energy Fuels **28**(5), 3420–3429 (2014)
8. J. Weber, D. Straub, A. Scholtissek, T.O. Brien, C. Olm, A. Konan, E.D. Huckaby, Chemical looping: Reactor experiments, modeling and simulation. In *NETL 2011 Workshop on Multiphase Flow Science*; Pittsburgh, PA (2011)
9. J.M. Parker, CFD model for the simulation of chemical looping combustion. Powder Technol. **265**, 47–53 (2014)

10. Z. Peng, E. Doroodchi, Y.A. Alghamdi, K. Shah, C. Luo, B. Moghtaderi, CFD-DEM simulation of solid circulation rate in the cold flow model of chemical looping systems. Chem. Eng. Res. Des. **95**, 262–280 (2015)
11. C. Luo, Z. Peng, E. Doroodchi, B. Moghtaderi, A three-dimensional hot flow model for simulating the alumina encapsulated NI-NIO methane-air CLC system based on the computational fluid dynamics-discrete element method. Fuel **224**, 388–400 (2018)
12. J. Ma, X. Tian, C. Wang, X. Chen, H. Zhao, Performance of a 50 kWth coal-Fuelled chemical looping combustor. Int. J. Greenh. Gas Control **75**(May), 98–106 (2018)
13. X. Chen, J. Ma, X. Tian, J. Wan, H. Zhao, CPFD simulation and optimization of a 50 KWth dual circulating fluidized bed reactor for chemical looping combustion of coal. Int. J. Greenh. Gas Control **90**, 102800 (2019)
14. X. Chen, J. Ma, X. Tian, Z. Xu, H. Zhao, Numerical investigation on the improvement of carbon conversion in a dual circulating fluidized bed reactor for chemical looping combustion of coal. Energy Fuel **33**, 12801–12813 (2019)
15. M. Su, H. Zhao, J. Ma, Computational fluid dynamics simulation for chemical looping combustion of coal in a dual circulation fluidized bed. Energy Convers. Manag. **105**, 1–12 (2015)
16. T. Pröll, K. Rupanovits, P. Kolbitsch, J. Bolhàr-Nordenkampf, H. Hofbauer, Cold flow model study on a dual circulating fluidized bed system for chemical looping processes. Chem. Eng. Technol. **32**(3), 418–424 (2009)
17. P. Kolbitsch, T. Pröll, H. Hofbauer, Modeling of a 120 KW chemical looping combustion reactor system using a Ni-based oxygen carrier. Chem. Eng. Sci. **64**(1), 99–108 (2009)
18. S. Wang, Y. Yang, H. Lu, P. Xu, L. Sun, Computational fluid dynamic simulation based cluster structures-dependent drag coefficient model in dual circulating fluidized beds of chemical looping combustion. Ind. Eng. Chem. Res. **51**(3), 1396–1412 (2012)
19. S. Wang, H. Lu, F. Zhao, G. Liu, CFD studies of dual circulating fluidized bed reactors for chemical looping combustion processes. Chem. Eng. J. **236**, 121–130 (2014)
20. S. Wang, J. Chen, H. Lu, G. Liu, L. Sun, Multi-scale simulation of chemical looping combustion in dual circulating fluidized bed. Appl. Energy **155**, 719–727 (2015)
21. C. Geng, W. Zhong, Y. Shao, D. Chen, B. Jin, Computational study of solid circulation in chemical-looping combustion reactor model. Powder Technol. **276**, 144–155 (2015)
22. Z. Reinking, H.S. Shim, K.J. Whitty, J.A.S. Lighty, Computational simulation of a 100 KW dual circulating fluidized bed reactor processing coal by chemical looping with oxygen uncoupling. Int. J. Greenh. Gas Control **90**, 102795 (2019)

Chapter 9
Partial-Loop CLC Simulations

9.1 2D Partial-Loop Simulation of a Circulating Fluidized Bed Reactor

Kruggel-Emden et al. [1] simulated the gas-solid flow and reactions in a high-velocity riser air reactor (AR) and a low-velocity bubbling fluidized bed fuel reactor (FR) by using a 2D two-fluid model. Figure 9.1 shows the outline of the combined framework with a buffer located in different places. The cyclone and siphon are neglected, and the exchange of the mass, momentum, heat, and species between the two reactors is realized by buffer and time-dependent boundary conditions. The buffer is used to reduce the fluctuations of the solid flow rate. When the solid inflow rate is smaller than the expected flow rate, a solid phase stream is introduced to the buffer. When the solid flow rate exceeds the target value, surplus particles are stored in the buffer. The buffer is regarded as an ideal mixer, and it is not an entity in the CFD model. In the simulation, Mn_3O_4 is used as the oxygen carrier and CH_4 is used as the fuel. The temperature distribution, gas composition distribution, and the degree of oxidation and reduction of the oxygen carrier in the reactor are studied.

Figure 9.2a displays the solid mass flow rates from and to the buffer which is located upstream from the bubbling fluidized bed FR. The mass flow $\dot{m}_{s,ox}$ from the outlet of the riser shows very obvious fluctuations because of the intensive gas-solid flow regime. The largest $\dot{m}_{s,ox}$ reaches the value of about 35 kg/s, which could be caused by some temporary solid clusters. To ensure the stable solid flow to the bubbling fluidized bed, the outlet solid flow rate from the buffer is set to 3.5275 kg/s. At first, the mass flows in the buffer are all from feed stream. After the time of 15 s, it is only occasionally necessary to add particles to the buffer from the feed stream as the particle circulation has been established in the system. Figure 9.2b displays the temperature of particles flowing into and out of the buffer as well as the particles inside the buffer. In the first 7 s, the solid temperature is 1223 K, as all the particles in the buffer are injected from the feed stream $\dot{m}_{s,feed}$ with constant temperature. After that, the solid temperature changes significantly with time due to the reaction of

R. K. Agarwal, Y. Shao, *Modeling and Simulation of Fluidized Bed Reactors for Chemical Looping Combustion*, https://doi.org/10.1007/978-3-031-11335-2_9

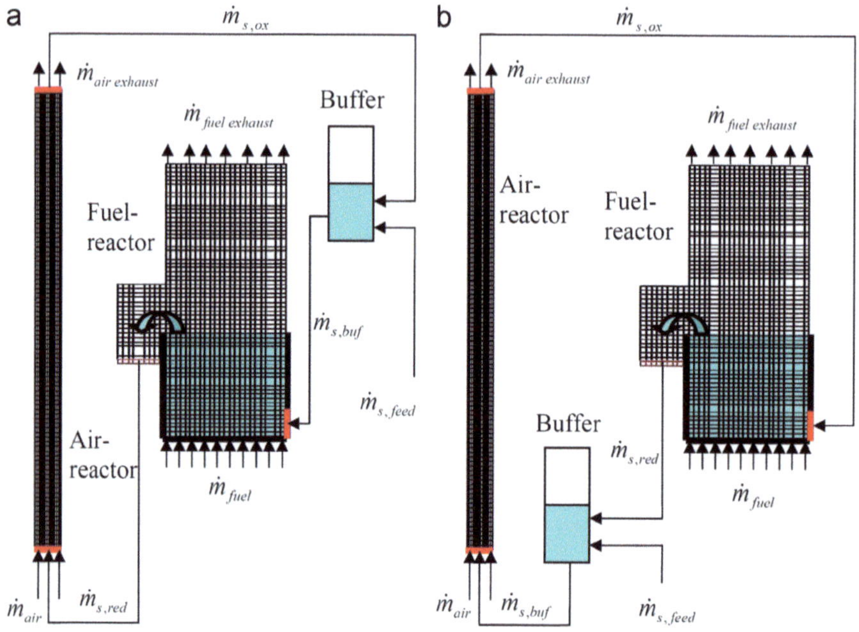

Fig. 9.1 Outline of the combined framework for partial-loop simulation developed by Kruggel-Emden et al. [1]: (a) the buffer is located upstream from the FR, and (b) the buffer is located upstream from the AR

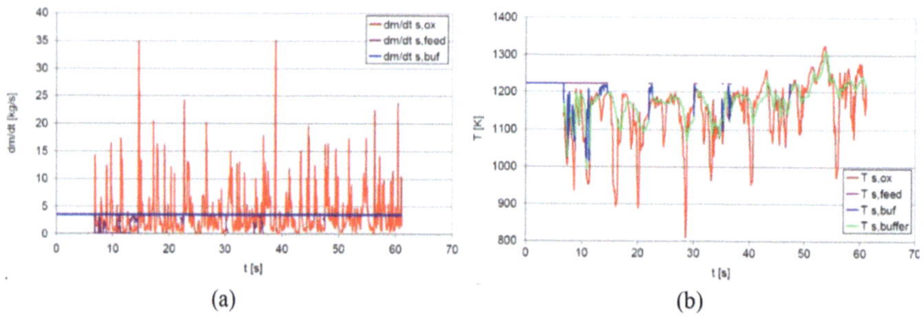

Fig. 9.2 (a) Mass flow into and out of the buffer and (b) solid temperature distribution where the buffer is placed upstream from the fuel reactor [1]

oxygen carriers. After the time of 15 s, the temperatures of particles from the riser outlet and buffer outlet agree well because only a small number of particles are needed to inject to the buffer from the feed stream.

When the buffer is located upstream of the bubbling fluidized bed fuel reactor, it can be seen from Fig. 9.2a that strong solid mass flow rate fluctuations occur at the outlet of the riser which further causes the intense temperature fluctuations. To prevent this situation, the buffer is placed upstream of the riser as shown in Fig. 9.1b to ensure constant solid flow rate injected to the riser. Figure 9.3a displays

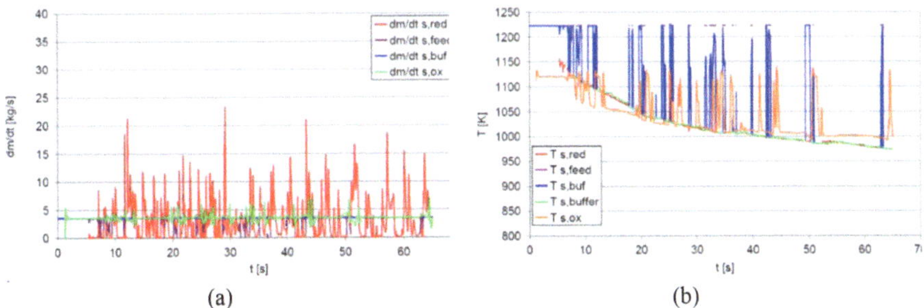

Fig. 9.3 (**a**) Mass flow into and out of the buffer, (**b**) solid temperature distribution where the buffer is placed upstream from the air reactor [1]

the solid flow rate fluctuations with time. Minor fluctuations are observed where the overall variations and the peak values are all smaller than those in Fig. 9.2a. Figure 9.2b displays the solid temperature distribution with time. In contrast to the base case, the temperature evolution is much smoother. Peaks are occasionally originating from fresh feed material entering the buffer.

9.2   2D Partial-Loop Simulation of a Dual-Loop Circulating Fluidized Bed Reactor

Zhang et al. [2] used a two-fluid model to perform a 2D cold-flow simulation of a dual-CFB CLC system. Figure 9.4a shows the experimental system in which the solid exchange is achieved by two cyclones, two loop seals, and a bottom lift. Figure 9.4b shows the simulation system where the cyclones, loop seals, and bottom lift are removed. The separation efficiency of the cyclones is assumed to be one. The bottom extraction/lift is replaced by an internal recirculation mechanism in order to keep the mass balance inside one reactor.

The fuel reactor and the air reactor are simulated independently using two sets of governing equations with different coordinates and parameters. The particles flowing out of the air reactor are injected at the bottom of the fuel reactor, while the same amount of particles in the fuel reactor outlet are fed into the air reactor inlet. In this way, the continuous solid circulation between the fuel reactor and air reactor is realized. Time-dependent inlet and outlet boundary conditions are used to connect the two reactors. At each time step, the simulations of two reactors are conducted separately, and the inlet solid flux is calculated from the solid flow rate monitored from the outlet of the other reactor.

Figure 9.5 shows the effects of the superficial gas velocity in FR on the average mass flow in the two reactor outlets. With the increase of the superficial gas velocity, the mass flow in the fuel reactor outlet increases, while the mass flow in the air reactor outlet almost remains constant. As the amount of exchanged solid between the reactors is determined by the lower solid flow rate, the increase of gas velocity in fuel reactor would enhance the solid exchange between the two reactors.

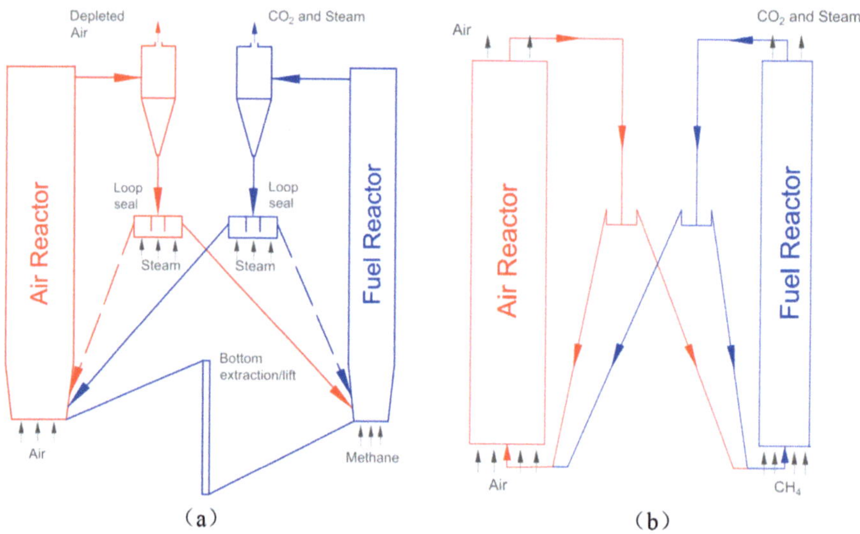

Fig. 9.4 (**a**) Experimental system [3] and (**b**) simulation system developed by Zhang et al. [2]

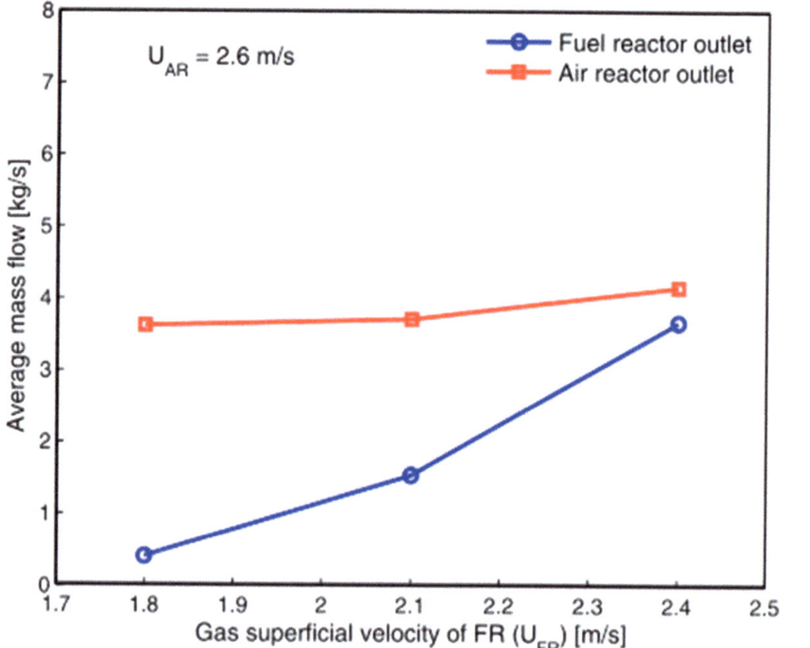

Fig. 9.5 Effect of the superficial gas velocity in FR on the average mass flow in the two reactor outlets [2]

9.3   3D Partial-Loop Simulation of a 120 kW Chemical Looping Combustion Pilot Plant

Hamidouche et al. [4] employed a two-fluid model to perform a 3D reactive simulation of a 120 kW CLC system operated at the Vienna University of Technology (TU Wien). Figure 9.6a displays the experimental configuration. The pilot plant employs the two circulating fluidized beds as the fuel reactor and air reactor, respectively, and they are connected to each other by upper and lower loop seals. Steam is injected to the two loop seals to avoid gas leakage between the two reactors. In the experiment, the natural gas is injected into the fuel reactor. The system also includes an internal solid circulation in the fuel reactor, and the returning of particles is realized by a loop seal which is also fluidized with steam.

Figure 9.6b shows the simulation geometry. The internal loop seal, upper loop seal, and cyclones are removed in the simulation, and they are replaced by appropriate boundary conditions, which can reduce the computational cost. In the experiment, gas is introduced into the bottom of the internal loop seal and upper loop seal to transport particles. In the simulation, the same amount of gas from the internal

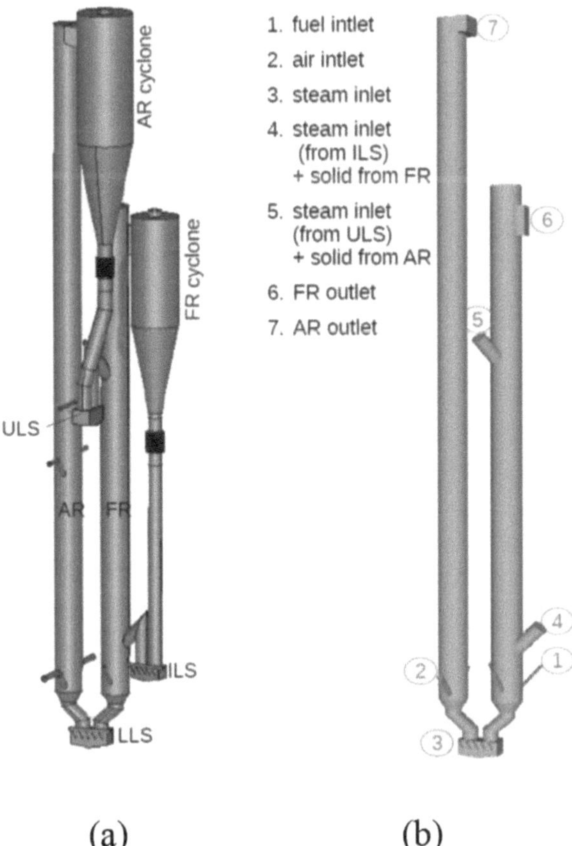

Fig. 9.6 (a) Experimental system [4] and (b) simulation system developed by Hamidouche et al. [4]

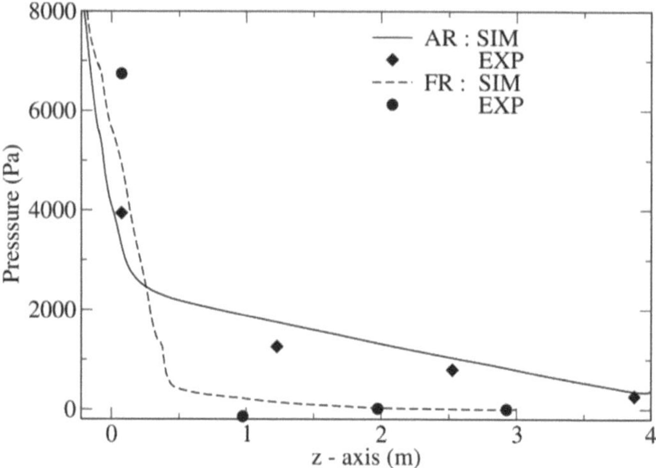

Fig. 9.7 Mean gas pressure-drop profiles [4]

loop seal inlet and upper loop seal inlet is fed into inlet 4 and inlet 5, respectively. At every time step, the solid mass flow rate at outlet 7 of the AR is counted, and the same amount of particles are introduced to the inlet 5 of the FR. The solid flow rate at the inlet 4 of the FR is equal to the solid flow rate at outlet 6. The pressures at the outlet 6 and outlet 7 are kept consistent with the experiment. As for the gas injection, the simulation and experiment are almost the same, and the only difference is that the pipes of the nozzles are a few centimeters inside the reactor in the experiment.

Figure 9.7 displays the comparison of the pressure profiles along the bed height in the two reactors between the simulation and experimental results. The results agree well in most regions but have a deviation at the bottom of the fuel reactor. This could be attributed to the fact that the steam is injected in the lower loop seal via the penetrating pipes which is not reproduced in the simulation.

References

1. H. Kruggel-Emden, S. Rickelt, F. Stepanek, A. Munjiza, Development and testing of an interconnected multiphase CFD-model for chemical looping combustion. Chem. Eng. Sci. **65**(16), 4732–4745 (2010)
2. Y. Zhang, Z. Chao, H.A. Jakobsen, Modelling and simulation of hydrodynamics in double loop circulating Fluidizedbed reactor for chemical looping combustion process. Powder Technol. **310**, 35–45 (2017)
3. A. Bischi, Ø. Langørgen, J.-X. Morin, J. Bakken, M. Ghorbaniyan, M. Bysveen, O. Bolland, Hydrodynamic viability of chemical looping processes by means of cold flow model investigation. Appl. Energy **97**, 201–216 (2012)
4. Z. Hamidouche, E. Masi, P. Fede, O. Simonin, K. Mayer, S. Penthor, Unsteady three-dimensional theoretical model and numerical simulation of a 120 KW chemical looping combustion pilot plant. Chem. Eng. Sci. **193**, 102–119 (2019)

Chapter 10
Binary Particle Bed Simulations in a Carbon Stripper

The use of solid coal fuel instead of gaseous fuels in chemical looping combustion introduces additional operational complexities. Unlike gaseous fuel, which is directly combusted by the oxygen carrier, the coal particles must first undergo a devolatilization process followed by a gasification reaction where the remaining char is reacted by the fluidizing gases consisting of recycled CO_2 and/or H_2O. The products of devolatilization and gasification are then combusted by the oxygen carrier. A typical CLC setup utilizes a cyclonic separator to isolate the oxygen carrier particles from the flue gases after the fuel reactor and the air reactor before transporting the solids between the reactors to continue to loop. Since char gasification is a slow process [1], unburnt char particles often remain in the flue stream of the fuel reactor. If these char particles are transported to the air reactor along with the oxygen carrier particles, the carbon capture efficiency of the CLC process would be reduced.

Several approaches have been proposed to prevent char particles from reaching the air reactor. One way is to provide sufficient residence time in the fuel reactor to ensure that the gasification reaction is complete. This can be achieved either by increasing the size of the reactor or by reducing the fluidizing gas velocity, but both options can impede the fluidization behavior of the bed, particularly in a spouted bed configuration. To avoid poor fluidization while still maintaining an increased residence time, a multi-staged fuel reaction concept was recently proposed and investigated [2]. A mass and energy balance study of the multi-staged fuel reaction setup conducted using Aspen Plus demonstrated that complete char conversion can be achieved by using multiple smaller fuel reactors in series such that any unburnt char in the system is burnt in subsequent fuel reactor stages before the solids are transported to the air reactor (see Chap. 3).

Figure 10.1 shows the differences in size between the particles of pulverized coal and a typical oxygen carrier (ilmenite) used in CLC operation. In fact, one reason of considering the spouted fluidized bed configuration is that it overcomes the limitation of a bubbling or fast fluidized bed to handle particles larger than a few hundred micrometers in diameter. Thus, one way of preventing the leakage of unburnt char

R. K. Agarwal, Y. Shao, *Modeling and Simulation of Fluidized Bed Reactors for Chemical Looping Combustion*, https://doi.org/10.1007/978-3-031-11335-2_10

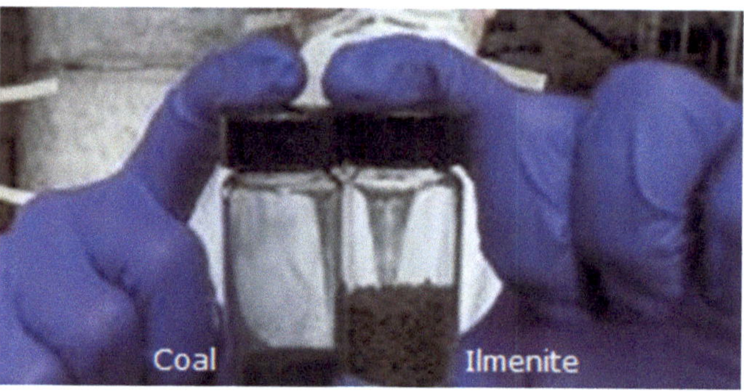

Fig. 10.1 Size difference between particles of coal and oxygen carrier

into the air reactor is to take advantage of the differences in size and density and hence the terminal velocity to separate the lighter char from the heavier oxygen carrier particles. Since char already has a lower density than the oxygen carrier, using pulverized coal particles smaller than or almost of the same size as the oxygen carrier particles should invariably lead to satisfactory separation results. The devolatilization and gasification processes that the coal undergoes further decrease the char particle size, enhancing the separation effect. The device that separates the char particles from the char and oxygen carrier mixture stream exiting the fuel reactor is known as a carbon stripper. The char particles from the carbon stripper can be returned to the fuel reactor to complete the gasification step while the oxygen carrier particles are transported to the air reactor to be regenerated.

By preventing the combustion of the unburnt char in atmospheric air, the carbon stripper also eliminates the formation of pollutants such as CO_2 and NO_X in the air reactor, as highlighted by Kramp et al. [3] and Mendiara et al. [4], and is deemed critical for CLC operation despite the increased hydrodynamic complexity associated with implementing the carbon stripper compared to increasing the residence time in the fuel reactor. It has been noted that for other fuels such as biomass where the increase in residence time required for the solid fuel is smaller, the direct increased residence time approach may be more competitive. In recent years, carbon strippers operating with fluidizing velocity in the range of 0.15–0.40 m/s have been incorporated into CLC experiments by Markström et al. [5], Ströhle et al. [6], Abad et al. [7], and Sun et al. [8] The results of these experiments indicated that the fluidization velocity should be increased further to increase the particle separation.

Later, Sun et al. [9] conducted cold-flow studies using a riser-based carbon stripper operating in the fast fluidized bed regime to investigate the effect of gas velocity on the separation ratio. The goal of Sun et al.'s design [9] was to achieve a high separation ratio to minimize the leakage of char particles into the air reactor with a low fluidizing gas velocity to keep operational costs low. However, the specific nature of the multiphase solid-gas flow inside the carbon stripper and how its geometry affects the design targets are not well understood from the experiment. In order to identify these relationships, a CFD-DEM-coupled simulation is described in this chapter for the carbon stripper consisting of a binary particle bed of coal and oxygen carrier particles and is validated against the experiment of Sun et al. [9] In future work, there is considerable scope to optimize the geometry of the carbon stripper to enhance the achievement of these design goals that can be addressed by integrating a multi-objective genetic algorithm with the CFD-DEM code.

10.1 Description of Experimental Setup

The carbon stripper used in the cold-flow experiment by Sun et al. [9] consisted of a riser, 4 m tall with a diameter of 0.7 m. A schematic of the experimental setup is given in Fig. 10.2. The solids mixture contained 95% ilmenite particles by mass and 5% plastic beads representing the unburnt char particles in the system.

The physical properties of ilmenite and plastic beads are listed in Table 10.1. The riser was fluidized from the bottom by air with the fluidizing velocity u_g in the range

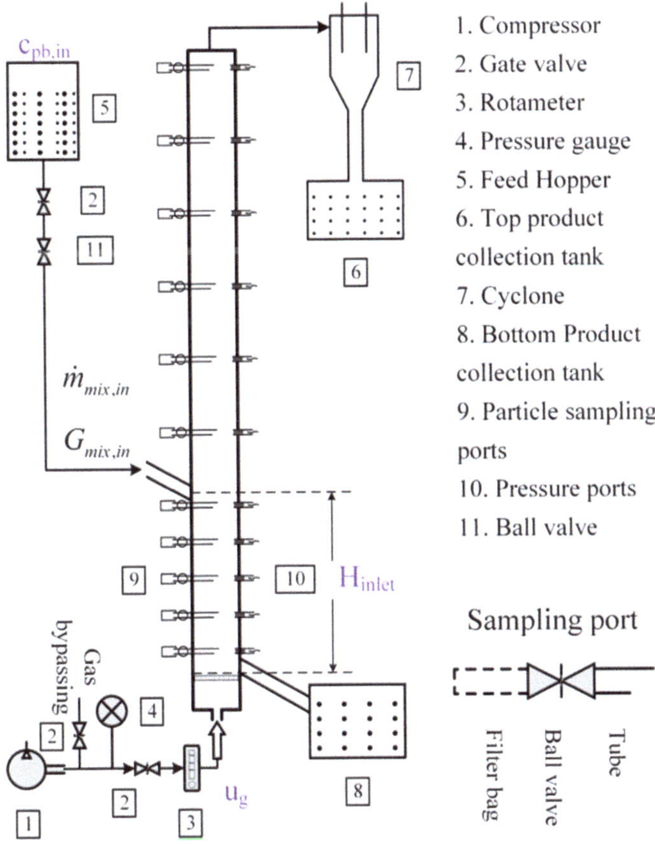

Fig. 10.2 Schematic of riser-based carbon stripper used by Sun et al. [9]

Table 10.1 Properties of ilmenite particles and plastic beads used by Sun et al. [9]

Particle	d_p (0.5) (µm)	ρ_p (kg/m^3)	u_t (m/s)
Ilmenite	257	4260	5.65
Plastic beads	94	960	0.39

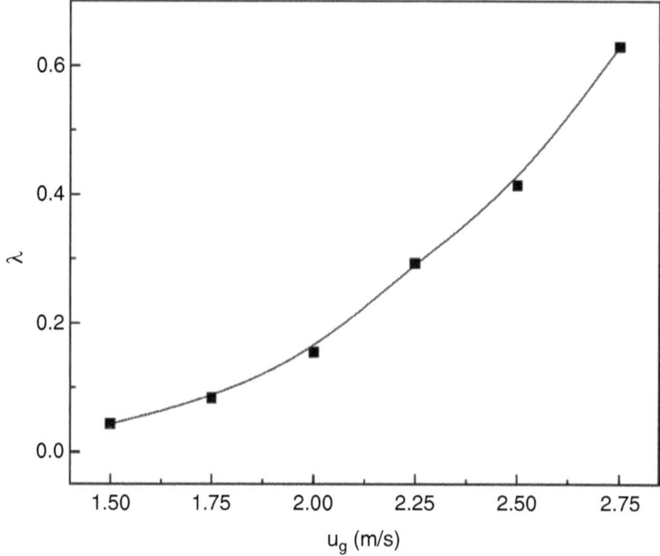

Fig. 10.3 Effect of fluidizing velocity on the separation efficiency λ for a solids mass flux of 12.2 kg/m^2-s^1

of 1.50–2.75 m/s increasing at 0.25 m/s intervals. u_g was selected to fall between the terminal velocities u_t for the ilmenite and plastic beads such that the plastic beads are carried out of the bed and exit the riser from the top into a tank while the ilmenite particles remain in the bed and collect at the bottom tank. The solids mixture is injected from the side of the riser at a height of $H_{inlet} = 1$ m above the bottom collection tank.

The separation ratio λ is defined as the mass of particles collected from top tank to the mass of particles collected from the top and bottom tanks combined, as given by.

$$\lambda = \frac{m_{mix,top}}{m_{mix,top} + m_{mix,btm}} \tag{10.1}$$

The concentration of plastic beads in each mixture sample in the experiment was determined by burning the mixture and measuring the change in weight. The plastic beads completely combusted to form CO_2 and H_2O, while the weight loss of the oxygen carrier was approximately 1% due to the reduction of ilmenite. λ is calculated for each u_g based on the experimental results and is plotted in Fig. 10.3 for a solids mixture feeding rate $G_{mix, in}$ of 12.2 kg/m^2-s; the same value of $G_{mix, in}$ is used in the simulations.

10.2   Computational Setup

The geometry used in the CFD-DEM simulation of the carbon stripper used by Sun et al. [9] employs the exact dimensions of the riser presented in Fig. 10.2. Since the solid flow from the feed hopper into the riser is of no consequence to the simulation, the solids inlet is simply modeled as a partial pipe. The top collection tank is also eliminated, and the solid flow at the top is measured directly at the riser outlet. The bottom collection tank is modeled as a simple closed boundary in order to ensure the accurate pressure boundary condition at the bottom of the riser; the solid flow into the bottom tank is measured at the surface between the riser and the tank. A structured grid is generated for all elements of the geometry and is shown in Fig. 10.4. The total number of cells is 51,884 in order to maintain a minimum cell volume greater than the particle (parcel) volume.

Given the small particle diameters of ilmenite and plastic beads used in the experiment (see Table 10.1), the number of particles in the system is very large. The parcel approach is employed to reduce the computational load with a parcel diameter of 0.002 m. Two solid injections are used corresponding to the ilmenite and

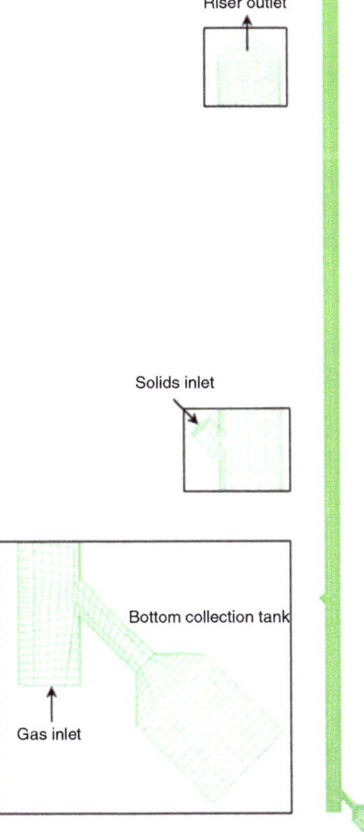

Fig. 10.4 Geometry with detailed views used in CFD-DEM simulation of the experiment of Sun et al. [9]

the plastic beads; the injection mass flow rates are calculated based on $G_{\text{mix, in}}$, the riser cross-section $A_r = \pi D_r^2 / 4$, and the concentration of plastic beads in the solid flow c_{pb}, as outlined below:

$$\dot{m}_{\text{mix,in}} = G_{\text{mix,in}} A_r \tag{10.2}$$

$$\dot{m}_{\text{ilm,in}} = \frac{\dot{m}_{\text{mix,in}}}{1 + c_{\text{pb}}} ; \quad \dot{m}_{\text{pb,in}} = c_{\text{pb}} \frac{\dot{m}_{\text{mix,in}}}{1 + c_{\text{pb}}} \tag{10.3}$$

The volumetric flow rates can be determined given the respective densities of the two materials and are used to determine the solid injection velocity:

$$q_{\text{ilm,in}} = \frac{\dot{m}_{\text{ilm,in}}}{\rho_{\text{ilm}}} ; \quad q_{\text{pb,in}} = \frac{\dot{m}_{\text{pb,in}}}{\rho_{\text{pb}}} \tag{10.4}$$

$$q_{\text{mix,in}} = q_{\text{ilm,in}} + q_{\text{pb,in}} \tag{10.5}$$

$$u_{\text{mix,in}} = \frac{q_{\text{mix,in}}}{A_r} \tag{10.6}$$

The soft sphere model is used for all the particle-particle and particle-wall collisions. In order to keep computing time low, the spring stiffness k_n is set at 5000 N/m to relax the minimum particle time step requirement. Bokkers [10] demonstrated that the results produced using this value of k_n are indistinguishable from those using larger values of k_n, which necessitate a smaller particle time step. The coefficient of restitution is set at 0.97. The numerical simulations are conducted using the phase-coupled SIMPLE scheme with second-order discretization in space and first order in time. The simulation cases modeled, and the key modeling parameters are summarized in Table 10.2.

Table 10.2 Key modeling parameters for binary particle bed simulation in the riser-based carbon stripper

Primary phase	Air
Discrete phase(s)	Ilmenite; plastic beads
Parcel diameter	0.2 m
Gas inlet fluidizing velocities	1.50, 1.75, 2.00, 2.25, 2.75 m/s
Solid injection velocity	$u_{\text{mix, in}} = 0.0034$ m/s
Solid injection flow rate	$\dot{m}_{\text{ilm,in}} = 0.044$ kg/s; $\dot{m}_{\text{pb,in}} = 0.0023$ kg/s
Outlet	Pressure outlet at atmospheric pressure
Drag law	Gidaspow [11]
Particle collision model	Soft-sphere model
Spring constant	5000 N/m
Coefficient of restitution	0.97
Friction coefficient	0.5
Time step size	Particle: 5×10^{-5} s; fluid: 5×10^{-4} s

10.3   Binary Particle Bed Simulation Results

Each CFD-DEM simulation of the binary particle bed in the riser-based carbon stripper is run for 20 s. The solid flow rate out of the riser outlet and into the bottom collection tank as well as the static pressure in the bed is recorded every 20 time steps (0.01 s). In the experiment of Sun et al. [9], after the initial development of fluidization caused by the solid injection, the pressure differences across sampling ports 1–3 and 8–11 shown in Fig. 10.2 stabilized after approximately 10 s. The static pressure at 2 mm above the inlet is used to verify the stable bed in the numerical simulation; the results for the $u_\mathrm{g} = 1.50$ m/s case are given in Fig. 10.5. It can be seen that the static pressure in the bed stabilizes after approximately 8 s. The final 10 s of simulation is used as the averaging interval for the solid flow rates in order to calculate the separation ratio λ, which is used to quantitatively validate the accuracy of the simulation against the experimental results [9]. To confirm that the averaging interval does not affect the value of λ, the simulation with $u_\mathrm{g} = 2.00$ m/s was run for 30 s, and the computed difference in λ was miniscule.

The development of solid flow into and out of the riser can be ascertained by examining the number of particles (parcels) of ilmenite and plastic beads held up in the riser after 20 s of simulation as shown in Table 10.3. As u_g increases, the number of ilmenite parcels in the riser increases. This is because the increased gas velocity

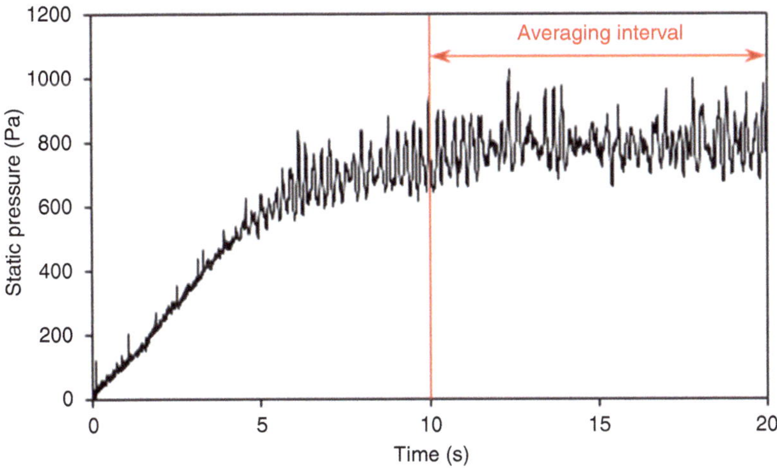

Fig. 10.5 Static pressure at 2 mm for binary particle bed simulation with fluidizing velocity $u_\mathrm{g} = 1.50$ m/s

Table 10.3 Number of parcels in the riser after 20 s of simulation for different fluidizing velocities

Particle	1.50 m/s	1.75 m/s	2.00 m/s	2.25 m/s	2.50 m/s	2.75 m/s
Ilmenite	17,267	26,138	34,064	41,044	45,741	14,051
Plastic beads	4882	3570	2403	1383	1211	1059

prevents the ilmenite from settling at the bottom of the riser and flowing into the bottom collection tank. However, this increase starts to diminish around $u_g = 2.50$ m/s, and for $u_g = 2.75$ m/s, the number of ilmenite particles in the riser decreases drastically. Although 2.50 m/s is still lower than the terminal velocity of ilmenite, the decreased hold-up suggests that at this velocity, the flow has sufficient energy to carry the particles out of the riser. On the other hand, the number of plastic beads in the riser steadily decrease as the fluidizing velocity increases. This is expected since the u_g/u_t ratio starts out at more than 1 at $u_g = 1.50$ m/s, and as it gets larger, the flow is able to carry the plastic beads out with greater ease.

For each case, the solid flow out of the riser outlet consists almost entirely of plastic beads with a few ilmenite particles, except at $u_g = 2.75$ m/s. On the other hand, the solid flow into the bottom collection tank is solely composed of ilmenite. This is expected given that the fluidizing velocities in each case lie between the terminal velocities of the plastic beads and ilmenite particles such that the fluid can carry the lighter plastic beads out of the bed but not the ilmenite particles. The flow rate of plastic beads out of the top riser outlet for different values of u_g is presented in Fig. 10.6. As u_g increases, the plastic beads reach the outlet faster because of a higher induced particle velocity, and the overall flow rate increases slightly until it stabilizes at a roughly constant value in each case equal to the injection flow rate of plastic beads; the plastic beads' flow rate into the bottom collection tank is nil. The transient fluctuations in the flow rate are due to the highly unsteady flow in the fast fluidization regime associated with the riser.

Similar plots are generated for the flow rate of ilmenite out of the top of the riser and into the bottom collection tank and are shown in Figs. 10.7 and 10.8, respectively. As mentioned above, the ilmenite flow rate out of the top outlet is limited to isolated particles up to $u_g = 2.25$ m/s. The ilmenite flow rate into the bottom collection tank decreases as u_g increases. The flow rate plots confirm the solid flow behavior suggested by the parcel hold-up numbers in Table 10.3.

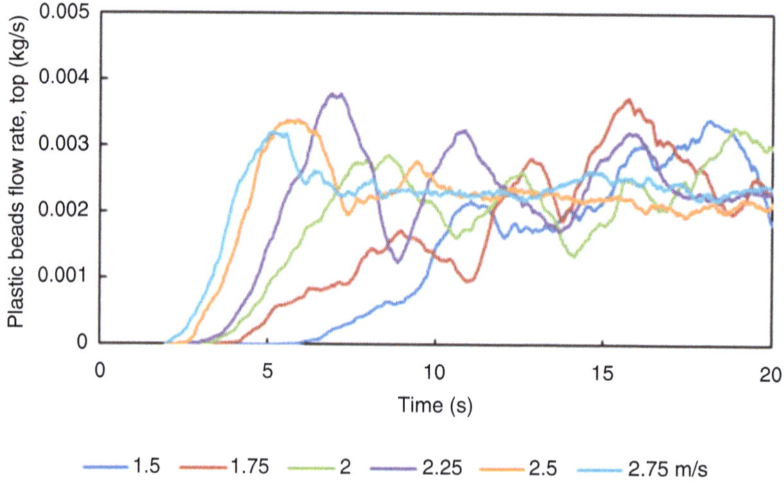

Fig. 10.6 Plastic bead flow rate out of top of the riser for different fluidizing velocities

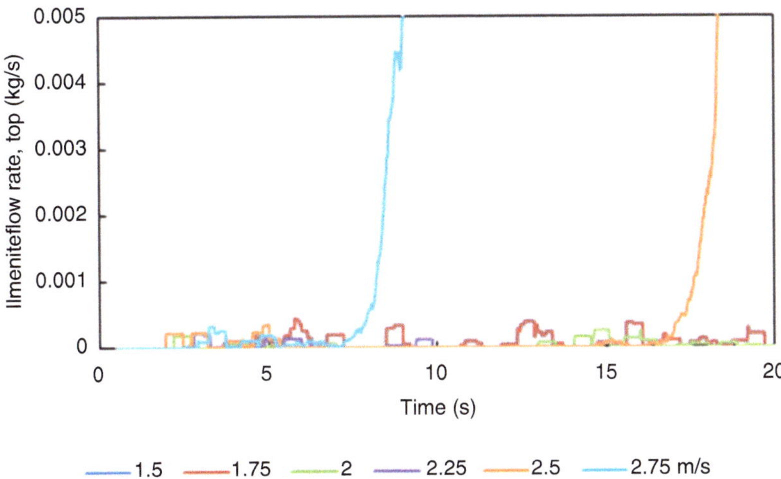

Fig. 10.7   Ilmenite flow rate out of top of the riser for different fluidizing velocities

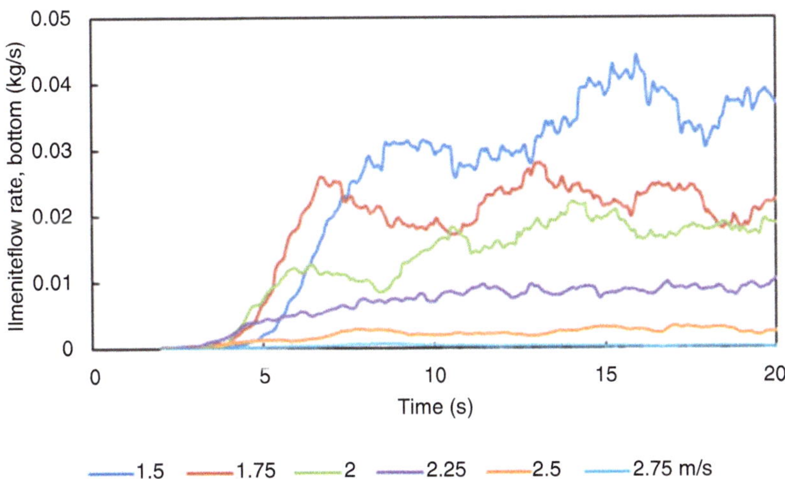

Fig. 10.8   Ilmenite flow rate into bottom collection tank for different fluidizing velocities

The flow rates of the plastic beads out of the top of the riser and the ilmenite into the bottom collection tank shown in Figs. 10.6 and 10.8, respectively, are used to compute the separation ratio λ according to Eq. (10.1). The values of λ for different fluidization velocities are plotted in Fig. 10.9. The values of λ in Fig. 10.9 are in excellent agreement with the experimental values presented in Fig. 10.3. Hence, the binary particle bed simulation conducted in this chapter can be considered to be a credible model for the experiment and can be employed to examine additional changes to the geometry and operating conditions and to investigate their effect on λ.

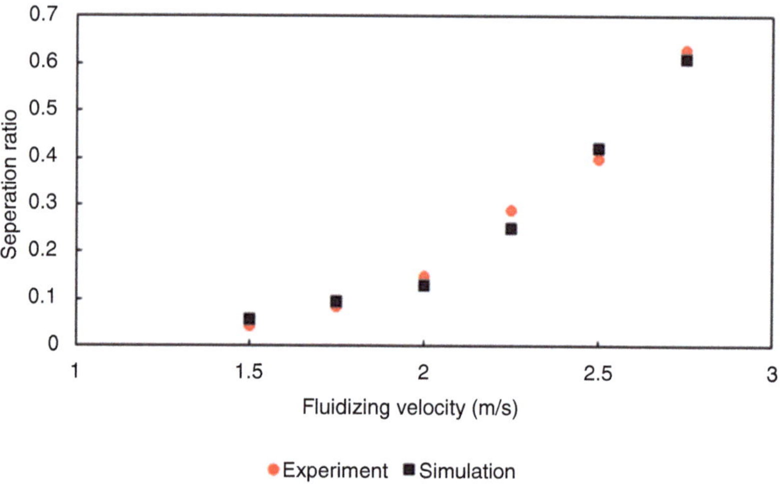

Fig. 10.9 Effect of fluidizing velocity on separation ratio λ for a solids mass flux of 12.2 kg/m²-s in the CFD-DEM simulation of the riser-based carbon stripper compared against the experiment of Sun et al. [9]

References

1. H. Leion, T. Mattisson, A. Lyngfelt, The use of petroleum coke as fuel in chemical-looping combustion. Fuel **86**(12–13), 1947–1958 (2007)
2. W.X. Meng, S. Banerjee, X. Zhang, R.K. Agarwal, Process simulation of multi-stage chemical-looping combustion using Aspen Plus. Energy **90**, 1869–1877 (2015)
3. M. Kramp, A. Thon, E. Hartge, S. Heinrich, J. Werther, Carbon stripping – A critical process step in chemical looping combustion of solid fuels. Chem. Eng. Technol. **35**(3), 497–507 (2012)
4. T. Mendiara, M.T. Izquierdo, A. Abad, L.F. De Diego, F. García-Labiano, P. Gayán, J. Adánez, Release of pollutant components in CLC of lignite. Int. J. Greenh. Gas Control **22**, 15–24 (2014)
5. P. Markström, C. Linderholm, A. Lyngfelt, Chemical-looping combustion of solid fuels–design and operation of a 100 KW unit with bituminous coal. Int. J. Greenh. Gas Control **15**, 150–162 (2013)
6. J. Ströhle, M. Orth, B. Epple, Design and operation of a 1 MWth chemical looping plant. Appl. Energy **113**, 1490–1495 (2014)
7. A. Abad, R. Pérez-Vega, L.F. de Diego, F. García-Labiano, P. Gayán, J. Adánez, Design and operation of a 50 KW_th chemical looping combustion (CLC) unit for solid fuels. Appl. Energy **157**, 295–303 (2015)
8. H. Sun, M. Cheng, D. Chen, L. Xu, Z. Li, N. Cai, Experimental study of a carbon stripper in solid fuel chemical looping combustion. Ind. Eng. Chem. Res. **54**(35), 8743–8753 (2015)
9. H. Sun, M. Cheng, Z. Li, N. Cai, Riser-based carbon stripper for coal-fueled chemical looping combustion. Ind. Eng. Chem. Res. **55**(8), 2381–2390 (2016)
10. G.A. Bokkers, Multi-level modelling of the hydrodynamics in gas phase polymerisation reactors. [PhD Thesis - Research UT, graduation UT, University of Twente]. University of Twente (2005)
11. D. Gidaspow, D. Gidaspow, Multiphase flow and fluidization: Continuum and kinetic theory descriptions. AICHE J. **42**(4), 1197–1198 (1994)

Chapter 11
Review of Simulations of Gas-Fueled and Solid-Fueled CLC Process

11.1 Gas-Fuel-Based CLC Process

In the gas-fueled CLC process, the gas fuels directly contact and react with solid oxygen carriers, and the reduction of oxygen carrier is a critical part of the CLC process which has been simulated by a number of researchers as given in Table 11.1. The shrinking core model (SCM) and nucleation and nuclei growth model (NNGM) are two of the most frequently used models to describe the reduction of the OCs [1]. In the CFD simulations of gas-fueled CLC process, the SCM is more widely used to calculate the conversion rate of the OCs.

The first gas-fueled CLC simulation was performed by Jung and Gamwo [3] in a bubbling fluidized bed fuel reactor. As shown in Table 11.1, in many gas-fueled CLC simulations, CH_4 is used as the fuel, and NiO is used as the oxygen carrier. The gas-solid reaction rate is higher in the emulsion phase due to a larger solid volume fraction; therefore a higher concentration of CH_4 is observed in the bubbles. The heterogeneous reaction rate decreases along the bed height. When the reaction reaches quasi-equilibrium, the concentrations of gas reactant and product show typical oscillations in the dense phase due to bubble passage, but the fluctuations in the freeboard region are less intense because of the absence of solid reactant [18]. Mahalatkar et al. [19] found that with increasing superficial gas velocity, the reactant concentration increases, and product concentration decreases at the outlet, mainly because the bubble frequency is higher, and even slugs occur at higher gas velocity. Wang et al. [17, 18] built a 3D two-fluid model to investigate the gas-fueled CLC process between the coal gas and CuO as OC and pointed out three effective ways to promote the gas fuel conversion including increasing the initial bed height, increasing the bed temperature, and increasing the operating pressure. Harichandan et al. [20] reported another two methods of increasing gas-solid reaction efficiency including decreasing the bed width and using particles with smaller sizes. Lin et al. [5] established a 2D two-fluid model to analyze the reaction of CH_4 with a Ni-Cu-based oxygen carrier. The reaction behaviors were investigated under different

© The Author(s), under exclusive license to Springer Nature Switzerland AG 2024
R. K. Agarwal, Y. Shao, *Modeling and Simulation of Fluidized Bed Reactors for Chemical Looping Combustion*, https://doi.org/10.1007/978-3-031-11335-2_11

Table 11.1 A summary of the key inputs in the CFD simulations of gas-fueled CLC process [2]

Fuel	Solid density (kg/m^3)	Solid diameter (mm)	Initial temperature (K)	Reaction kinetics model	Turbulence model	References
CH_4	3589	0.12	1223	SCM	$\kappa - \varepsilon$	Jung and Gamwo [3]
CH_4	3446	0.2	1223	SCM	–	Porrazzo et al. [4]
CH_4	2000	0.2	1223	SCM	–	Lin et al. [5]
CH_4	–	0.37	1073	SCM	–	Chen et al. [6]
CH_4	3416	0.135	1100–1200	SCM	$\kappa - \varepsilon$	Zhang et al. [7]
CH_4	–	0.1525	1223	SCM	$\kappa - \varepsilon$	Kruggel-Emden [8]
CH_4	1700	0.149	1000–1100	SCM	$\kappa - \varepsilon$	Zhang et al. [9]
CH_4	3416	0.135	1173 (FR), 1223 (AR)	SCM	–	Wang et al. [10]
CH_4	3446	0.138	800–900 (FR), 1000–1100 (AR)	SCM	–	Luo et al. [11]
CH_4	3416	0.135	1173 (FR), 1223 (AR)	SCM	–	Wang et al. [12]
CH_4	2470	0.2	1223 (AR, LS), 1153 (FR)	SCM	–	Ahmed and Lu [13]
H_2	2960	0.2–0.3	1123–1248	SCM	$\kappa - \varepsilon$	Deng et al. [14]
H_2	2960	0.275	1123	SCM	$\kappa - \varepsilon$	Deng et al. [15]
H_2	2960 ($CaSO_4$) 2800 (CaS)	0.275	1123	SCM, NNGM	$\kappa - \varepsilon$	Ben-Mansour et al. [16]
CO, H_2	1560	0.5	873–1123	SCM	$\kappa - \varepsilon$	Wang et al. [17]
CO, H_2	1560	0.5	1073	SCM	$\kappa - \varepsilon$	Wang et al. [18]

weight ratios of NiO/CuO. Results showed that with the addition of CuO to the Ni-based oxygen carrier, the zero emission of H_2 and CO can be achieved in the gas-fueled CLC process because of higher reactivity of CuO. Details of each case mentioned in Table 11.1 can be obtained from the corresponding reference.

Table 11.2 A summary of the key inputs in the CFD simulations of solid-fueled CLC process [2]

Fuel	Solid density (kg/m^3)	Solid diameter (mm)	Initial temperature (K)	Turbulence model	References
Coal	1333 (coal)	0.013–0.32 (OC) 0.05–0.15 (coal)	1273	–	Parker [22]
Chinese bituminous coal	3472 (OC)	0.18 (OC)	1223	–	Su et al. [23]
Indian coal	5300 (Fe_2O_3) 5100 (Fe_3O_4) 6310 (CuO) 6000 (Cu_2O) 950 (coal)	0.1 (OC) 0.1 (coal)	1223	$\kappa - \varepsilon$	Menon and Patnaikuni [24]
Bituminous coal	4500 (OC)	0.12 (OC)	1173	$\kappa - \varepsilon$	Alobaid et al. [25]
Bituminous coal	Not given	0.154 (OC) 3.1 (coal)	1173	$\kappa - \varepsilon$	Sharma et al. [26]
Shenhua bituminous coal	3650 (OC) 1346 (coal)	0.15–0.35 (OC) 0.1–0.3 (coal)	1273	–	Chen et al. [27, 28]
Biomass (pine sawdust)	4462 (OC) 800 (biomass)	0.2 (OC) 0.75 (biomass)	1188	–	Yin et al. [29]
Wyoming Powder River Basin (PRB) coal	4100 (OC) 1324 (coal)	0.1–0.2 (OC) 0.05–0.225 (coal)	973 (FR) 1173 (AR)	Large eddy simulation (LES)	Reinking et al. [30]

Reactions			**References**
$Coal \rightarrow C + CH_4 + CO + H_2O + CO_2$	$C + H_2O \leftrightarrow CO + H_2$	$C + CO_2 \leftrightarrow 2CO$	Parker [22]
$CO + H_2O \leftrightarrow CO_2 + H_2$	$4FeO + O_2 \rightarrow 2Fe_2O_3$	$4Fe_2O_3 + CH_4 \rightarrow 8FeO + CO_2 + 2H_2O$	
$Fe_2O_3 + H_2 \rightarrow 2FeO + H_2O$	$Fe_2O_3 + CO \rightarrow 2FeO + H_2O$		
$Coal + 0.0666CO_2 \rightarrow 0.4663Char + 0.05192CH_4 + 0.00554C_2H_6 + 0.1672CO + 0.0604H_2 + 0.0269H_2O + 0.01007Ash$			Su et al. [23]
$Char + CO_2 \rightarrow 2CO$	$Char + H_2O \rightarrow CO + H_2$	$H_2 + 3Fe_2O_3 \rightarrow 2Fe_3O_4 + H_2O$	
$C_2H_6 + 21Fe_2O_3 \rightarrow 14Fe_3O_4 + 3H_2O + 2CO_2$		$CO + 3Fe_2O_3 \rightarrow 2Fe_3O_4 + CO_2$	
$CH_4 + 12Fe_2O_3 \rightarrow 8Fe_3O_4 + 2H_2O + CO_2$	$CO + H_2O \rightarrow CO_2 + H_2$	$4Fe_3O_4 + O_2 \rightarrow 6Fe_2O_3$	
$Coal \rightarrow 0.008N_2 + 0.301H_2 + 0.201H_2O + 0.010CH_4 + 0.37C + 0.105Ash$			Menon and Patnaikuni [24]
$C + CO_2 \rightarrow 2CO$	$C + 2H_2O \rightarrow CO + 2H_2$	$3Fe_2O_3 + CO \rightarrow 2Fe_3O_4 + CO_2$	
$12Fe_2O_3 + CH_4 \rightarrow 8Fe_3O_4 + 2H_2O + CO_2$	$3Fe_2O_3 + H_2 \rightarrow 2Fe_3O_4 + H_2O$	$2CuO + CO \rightarrow CO_2 + Cu_2O$	
$8CuO + CH_4 \rightarrow CO_2 + 4Cu_2O + 2H_2O$	$2CuO + H_2 \rightarrow Cu_2O + H_2O$		

(continued)

Table 11.2 (continued)

Fuel	Solid density (kg/m^3)	Solid diameter (mm)	Initial temperature (K)	Turbulence model	References
$Coal \rightarrow C + CH_4 + CO + H_2O + N_2$	$C + H_2O \rightarrow H_2 + CO$		$C + CO_2 \rightarrow 2CO$		Alobaid et al. [25]
$CH_4 + 4Fe_2O_3 + 8TiO_2 \rightarrow 2H_2O + CO_2 + 8FeTiO_3$			$H_2 + Fe_2O_3 + 2TiO_2 \rightarrow H_2O + 2FeTiO_3$		
$CO + Fe_2O_3 + 2TiO_2 \rightarrow CO_2 + 2FeTiO_3$			$C_3H_8 + 5O_2 \rightarrow 3CO_2 + 4H_2O$		
$Raw\ Coal \rightarrow C + CH_4 + CO_2 + H_2 + CO + H_2O + N_2$			$C + CO_2 \rightarrow 2CO$		Sharma et al. [26]
$C + H_2O \rightarrow H_2 + CO$	$CO + H_2O \rightarrow CO_2 + H_2$		$H_2 + Fe_2O_3 + 2TiO_2 \rightarrow H_2O + 2FeTiO_3$		
$CH_4 + 4Fe_2O_3 + 8TiO_2 \rightarrow 2H_2O + CO_2 + 8FeTiO_3$			$CO + Fe_2O_3 + 2TiO_2 \rightarrow CO_2 + 2FeTiO_3$		
$Coal + 0.001748H_2O \rightarrow 0.4846Char + 0.09427CO + 0.03989CO_2 + 0.02236CH_4 + 0.01723H_2 + 0.3434Ash$					Chen et al. [27, 28]
$Char + CO_2 \leftrightarrow 2CO$	$Char + H_2O \leftrightarrow CO + H_2$		$CO + 3Fe_2O_3 \rightarrow 2Fe_3O_4 + CO_2$		
$CH_4 + 12Fe_2O_3 \rightarrow 8Fe_3O_4 + 2H_2O + CO_2$	$H_2 + 3Fe_2O_3 \rightarrow 2Fe_3O_4 + H_2O$		$4Fe_3O_4 + O_2 \rightarrow 6Fe_2O_3$		
$Char + O_2 \rightarrow CO_2$	$CO + H_2O \rightarrow CO_2 + H_2$				
$Biomass \rightarrow CO + H_2 + CH_4 + CO_2 + H_2O + Char + Ash$					Yin et al. [29]
$C + CO_2 \rightarrow aCO$	$C + H_2O \rightarrow H_2 + CO$		$CO + H_2O \leftrightarrow H_2 + CO_2$		
$12Fe_2O_3 + CH_4 \rightarrow 8Fe_3O_4 + CO_2 + 2H_2O$	$3Fe_2O_3 + CO \rightarrow 2Fe_3O_4 + CO_2$		$3Fe_2O_3 + H_2 \rightarrow 2Fe_3O_4 + H_2O$		
$Coal \rightarrow C + CH_4 + CO + H_2 + H_2O + Ash$	$CO + 0.5O_2 \rightarrow CO_2$		$CO + H_2O \leftrightarrow CO_2 + H_2$		Reinking et al. [30]
$H_2 + 0.5O_2 \rightarrow CO$	$CH_4 + 2O_2 \rightarrow CO_2 + 2H_2O$		$C + 0.5O_2 \rightarrow CO$		
$C + O_2 \rightarrow CO_2$	$2CuO \rightarrow 0.5O_2 + Cu_2O$		$Cu_2O + 0.5O_2 \rightarrow 2CO$		
$C + H_2O \leftrightarrow CO + H_2$	$C + CO_2 \rightarrow 2CO$		$C + 2H_2 \leftrightarrow CH_4$		

11.2 Solid-Fuel-Based CLC Process

As mentioned in Chap. 2, the solid-fuel-based CLC process can be categorized into two types: in situ gasification chemical looping combustion (iG-CLC) process and chemical looping with oxygen uncoupling (CLOU) process. In the iG-CLC process, the solid fuels are gasified by CO_2 or H_2O. The oxygen carrier subsequently reacts with the gasification products and volatiles released by the solid fuels. In the CLOU process, the oxygen carrier releases gaseous oxygen which directly reacts with the char and volatiles leading to faster reaction than gasification. Table 11.2 gives some typical examples of iG-CLC and CLOU processes with different OCs and solid fuels. In a reactive model, the thermal and transport properties including gas viscosity, thermal conductivity, and heat capacity are all functions of the temperature and the compositions of the chemical species [21].

The first CFD simulation of iG-CLC process was performed by Mahalatkar et al. [31] with coal as the solid fuel and Fe_2O_3 as the oxygen carrier. The char gasification reaction is slowest, and it is the rate-controlling step. The carbon conversion can be improved by increasing the operating temperature and gasifying agent concentration. Another effective way to promote char gasification is elevating the operating pressure, which has been validated by Wang et al. [32] in a pressurized circulating fluidized bed riser fuel reactor. Oxygen carrier plays an important role in the iG-CLC process, and the effects of CuO and Fe_2O_3 and a mixture of the two materials were analyzed by Menon et al. [24] When the ash is not considered, CuO exhibits superior performance, which is consistent with the thermodynamic analysis. When the ash is included, the CuO mixture with oxygen carrier shows better performance. However, the performance of mixture of CuO and oxygen carrier is unstable throughout the temperature range, and Fe_2O_3 is preferable to ensure the stable running of the reactor. The iG-CLC processes were also studied by Alobaid et al. [25], Sharma et al. [26], and Yin et al [29] among others.

References

1. H. Kruggel-Emden, F. Stepanek, A. Munjiza, A study on the role of reaction modeling in multiphase CFD-based simulations of chemical looping combustion. Oil Gas Sci. Technol. **66**(2), 313–331 (2011)
2. Y. Shao, R.K. Agarwal, X. Wang, B. Jin, Review of computational fluid dynamics studies on chemical looping combustion. J. Energy Resour. Technol. Trans. ASME **143**(8) (2021)
3. J. Jung, I.K. Gamwo, Multiphase CFD-based models for chemical looping combustion process: Fuel reactor modeling. Powder Technol. **183**(3), 401–409 (2008)
4. R. Porrazzo, G. White, R. Ocone, Fuel reactor modelling for chemical looping combustion: From micro-scale to macro-scale. Fuel **175**, 87–98 (2016)
5. J. Lin, K. Luo, L. Sun, S. Wang, C. Hu, J. Fan, Numerical investigation of nickel-copper oxygen carriers in chemical-looping combustion process with zero emission of CO and H_2. Energy Fuels **33**(11), 12096–12105 (2019)

6. L. Chen, X. Yang, G. Li, X. Li, C. Snape, Prediction of bubble fluidisation during chemical looping combustion using CFD simulation. Comput. Chem. Eng. **99**, 82–95 (2017)
7. Y. Zhang, Z. Chao, H.A. Jakobsen, Modelling and simulation of chemical looping combustion process in a double loop circulating fluidized bed reactor. Chem. Eng. J. **320**, 271–282 (2017)
8. H. Kruggel-emden, S. Rickelt, F. Stepanek, A. Munjiza, Development and testing of an interconnected multiphase CFD-model for chemical looping combustion. Chem. Eng. Sci. **65**(16), 4732–4745 (2010)
9. Y. Zhang, Ø. Langørgen, I. Saanum, Z. Chao, H.A. Jakobsen, Modeling and simulation of chemical looping combustion using a copper-based oxygen carrier in a double-loop circulating fluidized bed reactor system. Ind. Eng. Chem. Res. **56**(50), 14754–14765 (2017)
10. S. Wang, J. Chen, H. Lu, G. Liu, L. Sun, Multi-scale simulation of chemical looping combustion in dual circulating fluidized bed. Appl. Energy **155**, 719–727 (2015)
11. C. Luo, Z. Peng, E. Doroodchi, B. Moghtaderi, A three-dimensional hot flow model for simulating the alumina encapsulated NI-NIO methane-air CLC system based on the computational fluid dynamics-discrete element method. Fuel **224**, 388–400 (2018)
12. S. Wang, H. Lu, F. Zhao, G. Liu, CFD studies of dual circulating fluidized bed reactors for chemical looping combustion processes. Chem. Eng. J. **236**, 121–130 (2014)
13. A. Bougamra, L. Huilin, Modeling of chemical looping combustion of methane using a Ni-based oxygen carrier. Energy Fuels **28**(5), 3420–3429 (2014)
14. Z. Deng, R. Xiao, B. Jin, H.H. Qilei Song, Multiphase CFD modeling for a chemical looping combustion process (fuel reactor). Chem. Eng. Technol. **31**(12), 1754–1766 (2008)
15. Z. Deng, R. Xiao, B. Jin, Q. Song, Numerical simulation of chemical looping combustion process with $CaSO_4$ oxygen carrier. Int. J. Greenh. Gas Control **3**(4), 368–375 (2009)
16. R. Ben-Mansour, H. Li, M.A. Habib, Effects of oxygen carrier mole fraction, velocity distribution on conversion performance using an experimentally validated mathematical model of a CLC fuel reactor. Appl. Energy **208**, 803–819 (2017)
17. X. Wang, B. Jin, Y. Zhang, W. Zhong, S. Yin, Multiphase computational fluid dynamics (CFD) modeling of chemical looping combustion using a CuO/Al_2O_3 oxygen carrier: Effect of operating conditions on coal gas combustion. Energy Fuels **25**(8), 3815–3824 (2011)
18. X. Wang, B. Jin, W. Zhong, Y. Zhang, M. Song, Three-dimensional simulation of a coal gas fueled chemical looping combustion process. Int. J. Greenh. Gas Control **5**(6), 1498–1506 (2011)
19. K. Mahalatkar, J. Kuhlman, E.D. Huckaby, T.O. Brien, Computational fluid dynamic simulations of chemical looping fuel reactors utilizing gaseous fuels. Chem. Eng. Sci. **66**(3), 469–479 (2011)
20. A.B. Harichandan, T. Shamim, CFD analysis of bubble hydrodynamics in a fuel reactor for a hydrogen-fueled chemical looping combustion system. Energy Convers. Manag. **86**, 1010–1022 (2014)
21. W. Shuai, Y. Yunchao, L. Huilin, W. Jiaxing, X. Pengfei, L. Guodong, Hydrodynamic simulation of fuel-reactor in chemical looping combustion process. Chem. Eng. Res. Des. **89**(9), 1501–1510 (2011)
22. J.M. Parker, CFD model for the simulation of chemical looping combustion. Powder Technol. **265**, 47–53 (2014)
23. M. Su, H. Zhao, J. Ma, Computational fluid dynamics simulation for chemical looping combustion of coal in a dual circulation fluidized bed. Energy Convers. Manag. **105**, 1–12 (2015)
24. K.G. Menon, V.S. Patnaikuni, CFD simulation of fuel reactor for chemical looping combustion of Indian coal. Fuel **203**, 90–101 (2017)
25. F. Alobaid, P. Ohlemüller, J. Ströhle, B. Epple, Extended Euler–Euler model for the simulation of a 1 MW_{Th} chemical–looping pilot plant. Energy **93**, 2395–2405 (2015)
26. R. Sharma, J. May, F. Alobaid, P. Ohlemüller, J. Ströhle, B. Epple, Euler-Euler CFD simulation of the fuel reactor of a 1 MW_{Th} chemical-looping pilot plant: Influence of the drag models and specularity coefficient. Fuel **200**, 435–446 (2017)

27. X. Chen, J. Ma, X. Tian, J. Wan, H. Zhao, CPFD simulation and optimization of a 50 KWth dual circulating fluidized bed reactor for chemical looping combustion of coal. Int. J. Greenh. Gas Control **90**, 102800 (2019)
28. X. Chen, J. Ma, X. Tian, Z. Xu, H. Zhao, Numerical investigation on the improvement of carbon conversion in a dual circulating fluidized bed reactor for chemical looping combustion of coal. Energy Fuel **33**, 12801–12813 (2019)
29. W. Yin, S. Wang, K. Zhang, Y. He, Numerical investigation of in situ gasification chemical looping combustion of biomass in a fluidized bed reactor. Renew. Energy **151**, 216–225 (2020)
30. Z. Reinking, H.S. Shim, K.J. Whitty, J.A.S. Lighty, Computational simulation of a 100 KW dual circulating fluidized bed reactor processing coal by chemical looping with oxygen uncoupling. Int. J. Greenh. Gas Control **90**, 102795 (2019)
31. K. Mahalatkar, J. Kuhlman, E.D. Huckaby, T. O'Brien, CFD simulation of a chemical-looping fuel reactor utilizing solid fuel. Chem. Eng. Sci. **66**(16), 3617–3627 (2011)
32. X. Wang, B. Jin, Y. Zhang, Y. Zhang, X. Liu, Three dimensional modeling of a coal-fired chemical looping combustion process in the circulating fluidized bed fuel reactor. Energy Fuels **27**(4), 2173–2184 (2013)

Chapter 12
Scaling Laws for CFD-DEM Simulations of CLC

12.1 Parcels of Particles

For a typical CLC system, the computational cost of tracking each individual particle is prohibitive. One simple approach available in ANSYS Fluent [1] for reducing the computing load is to divide the particles into clusters called parcels. The motion of each parcel is determined as a whole by considering it as a single representative particle [1]. Parcel collisions are evaluated in the same manner as described in the DEM approach for a single particle in Chap. 6; however, the mass of the entire parcel is considered and not just that of a single representative particle. The parcel diameter is that of a sphere whose volume is the sum of the volumes of its constituent particles. Hence, specifying a parcel diameter equal to twice the particle diameter leads to a reduction in the number of individual particles tracked by the DEM solver by a factor of eight, with an even larger decrease in the number of collisions.

12.2 Scaling Methodology and Law of Glicksman et al. [2]

The parcel approach provides a good starting point in reducing the computational cost of the coupled CFD-DEM simulation. More robust scaling methodologies can be derived based on the principles of dynamic similarity. By nondimensionalizing the governing equations of multiphase flow, Glicksman [2] determined the controlling nondimensional parameters for gas-solid flows. One equation of note is the Ergun equation, which predicts the gas-solid momentum exchange coefficient considering both viscous and inertial effects:

R. K. Agarwal, Y. Shao, *Modeling and Simulation of Fluidized Bed Reactors for Chemical Looping Combustion*, https://doi.org/10.1007/978-3-031-11335-2_12

$$\frac{\Delta p}{L} = 1.75\frac{\rho_{\mathrm{f}}(1-\alpha_{\mathrm{f}})u_0^2}{d_{\mathrm{p}}\alpha_{\mathrm{f}}^3} + 150\frac{\mu_{\mathrm{f}}(1-\alpha_{\mathrm{f}})^2 u_0}{d_{\mathrm{p}}^2\alpha_{\mathrm{f}}^3} \tag{12.1}$$

The full set of scaling parameters was simplified by Glicksman et al. [2] by isolating the viscous and inertial terms on the right-hand side of the Ergun equation. The simplified parameters hold exactly at both low and high values of Re_{p} (i.e., for both viscosity-dominated and inertia-dominated gas-solid flows) and are reasonably accurate for the entire range of conditions where the Ergun equation is applicable for obtaining the interphase momentum exchange. They are particularly applicable to a spouted fluidized bed because the relatively large particle diameters and high fluidization velocity place such a system in the inertia-dominated regime. Glicksman et al. [2] proposed a scaling methodology where the dimensions of the fluidized bed are reduced and the superficial velocity u_0 is adjusted accordingly to maintain the same Froude number, $Fr(=u_0/\sqrt{gL})$. Assuming a geometry scale r with the subscripts ex and sc representing the exact and scaled systems, respectively, the square of the Froude number can be written as

$$\frac{u_{0,\mathrm{ex}}^2}{gL_{\mathrm{ex}}} = \frac{u_{0,\mathrm{sc}}^2}{gL_{\mathrm{sc}}} \Rightarrow \frac{u_{0,\mathrm{ex}}}{u_{0,\mathrm{sc}}} = \left(\frac{L_{\mathrm{ex}}}{L_{\mathrm{sc}}}\right)^{1/2} = r^{1/2} \tag{12.2}$$

In turn, the minimum fluidization velocity u_{mf} is adjusted by reducing the particle diameter to hold u_0/u_{mf} constant. By definition, the minimum fluidization velocity occurs when the pressure drop in the Ergun equation is equal to the gravitational force of the particle bed, as given by

$$g(\rho_{\mathrm{p}} - \rho_{\mathrm{f}}) = 1.75\frac{\rho_{\mathrm{f}}u_{\mathrm{mf}}^2}{d_{\mathrm{p}}\alpha_{\mathrm{f}}^3} + 150\frac{\mu_{\mathrm{f}}(1-\alpha_{\mathrm{f}})u_{\mathrm{mf}}}{d_{\mathrm{p}}^2\alpha_{\mathrm{f}}^3} \tag{12.3}$$

After calculating the minimum fluidization velocity for the original model, the particle scaling factor n can be found by substituting for the minimum fluidization velocity and rearranging to solve the quadratic equation for n given as

$$g(\rho_{\mathrm{p}} - \rho_{\mathrm{f}})n^2 - \frac{1.75\rho_{\mathrm{f}}(1-\alpha_{\mathrm{f}})\left(r^{1/2}u_{\mathrm{mf,ex}}^2\right)}{d_{\mathrm{p,ex}}\alpha_{\mathrm{f}}^3}n - \frac{150\mu_{\mathrm{f}}(1-\alpha_{\mathrm{f}})^2\left(r^{1/2}u_{\mathrm{mf,ex}}\right)}{d_{\mathrm{p,ex}}^2\alpha_{\mathrm{f}}^3} = 0 \tag{12.4}$$

Glicksman et al. [2] demonstrated the utility of their scaling methodology by considering 1/4 and 1/16 scale models of an experimental reactor and matching the solid fraction profiles across the systems. In the context of CFD-DEM, the simplified scaling methodology of Glicksman et al. [2] holds promise because the reduction in particle diameter is smaller than the geometry scale used, allowing for a reduction in the total number of particles required in the system to maintain the same bed height. Given the properties of glass beads and air and a geometry scale of 1/4, the particle diameter must be scaled by 0.62, resulting in a reduction in the number of particles by a factor of around 15 (= (0.62/0.25) [3]).

12.3 Scaling Methodology and Law of Link et al. [3]

The scaling approach of Glicksman et al. [2] was derived for scaling experimental fluidized beds to reduce cost; its applicability to CFD-DEM simulations due to the reduction in the number of particles is coincidental. It should be noted that the particle Reynolds number, Re_p, and the Archimedes number, Ar (the ratio of gravitational forces to inertial forces), for the scaled system according to Glicksman et al. [2] are roughly equal to 0.34 and 0.12 times their exact values for the 1/4 and 1/16 scaled models, respectively, in the limit of very small particles; for the larger particles, the difference is even greater and varies with particle density. The ratio of the superficial velocity to the terminal velocity of the particles u_0/u_t is also significantly decreased. These are important parameters for accurately capturing the fluidization behavior for a spouted bed that cannot be matched by the Glicksman's scaling law.

Link et al. [3] proposed a scaling methodology for CFD-DEM simulation that utilizes the ability to change the physical properties of the materials in a computational model, which is not possible in an experiment, in order to maintain the same Re_p and Ar in the scaled model. By keeping these two parameters the same, the u_{mf} is also held constant. Unlike Glicksman et al.'s approach, the scaling approach proposed by Link et al. [3], only scales up the particle diameter while retaining the original geometry. Hence, Fr remains constant without changing u_0, and the other scaling parameter used by Glicksman et al. [2] and u_0/u_{mf} are automatically matched. By increasing the particle diameter by a factor of n, the number of particles can be reduced by a factor of n^3 while maintaining the same particle volume. To maintain the same particle Reynolds number, the dynamic viscosity for the gas phase in the scaled simulation is defined as

$$\text{Re}_{p,sc} = \text{Re}_{p,ex} \Rightarrow \mu_{f,sc} = \frac{d_{p,sc}}{d_{p,ex}} \mu_{f,ex} = n\mu_{f,ex} \tag{12.5}$$

given that the fluid density does not change. In the same vein, the Archimedes number can be held constant by setting a new particle density for the scaled simulation:

$$\text{Ar} = \frac{gd_p^3}{\mu_f^2}(\rho_p - \rho_f); \text{Ar}_{sc} = \text{Ar}_{ex} \Rightarrow \rho_{p,2} = \frac{(\rho_{p,1} - \rho_{f,1})}{n} + \rho_{f,2} \tag{12.6}$$

12.4 Scaling Methodology and Law of Banerjee and Agarwal

The scaling law of Link et al. [3] addresses the shortcoming of the simplified Glicksman et al.'s scaling law [2] by maintaining the same $\mathrm{Re_p}$ and Ar between the original and scaled models as well as by maintaining the same Fr and u_0/u_{mf}. Unfortunately, the ratio u_0/u_t, a crucial parameter in defining the fluidization behavior in a fluidized bed, is reduced to 0.707 of its original value, which can be expected to reduce the spout velocity of the particles in the scaled system. To rectify this, a new scaling methodology was proposed by Banerjee and Agarwal [4] to keep u_0/u_t constant in addition to every other nondimensional parameter used in the other two scaling methodologies.

Hence, the final set of scaling parameters used in this scaling approach are Fr and u_0/u_{mf} from the simplified scaling law of Glicksman et al. [2], $\mathrm{Re_p}$ and Ar from the scaling law of Link et al. [3], and u_0/u_t. First, the geometry is scaled by a factor r and u_0 by a factor $r^{1/2}$ to keep Fr constant. Similar to Link et al. [3], the physical properties of the materials are changed to match certain nondimensional parameters independently while not affecting others. In this case, the fluid density is changed depending on the particle scaling factor n (yet to be determined) to keep $\mathrm{Re_p}$ constant given the change in u_0; therefore

$$\mathrm{Re}_{\mathrm{p,sc}} = \mathrm{Re}_{\mathrm{p,ex}} \Rightarrow \rho_{\mathrm{f,sc}} = \rho_{\mathrm{f,ex}} \frac{d_{\mathrm{p,ex}}}{d_{\mathrm{p,sc}}} \frac{u_{0,\mathrm{ex}}}{u_{0,\mathrm{sc}}} = \frac{\rho_{\mathrm{f,ex}}}{nr^{1/2}} \tag{12.7}$$

For typical values of $\mathrm{Re_p}$ for a spouted fluidized bed, the terminal velocity is defined as

$$u_t^2 = 3 \frac{(\rho_{\mathrm{p}} - \rho_{\mathrm{f}})g d_{\mathrm{p}}}{\rho_{\mathrm{f}}} \tag{12.8}$$

After calculating the terminal velocity for the original model, one can determine n by substituting for the scaled fluid density and terminal velocity and rearranging to solve the quadratic equation for n given as

$$\rho_{\mathrm{p}} n^2 - \frac{\rho_{\mathrm{f,ex}}}{r^{1/2}} n - \frac{1}{3} \frac{\rho_{\mathrm{f,ex}} u_{\mathrm{t,ex}}^2 r^{1/2}}{g d_{\mathrm{p,ex}}} = 0 \tag{12.9}$$

Using the particle scaling factor and the corresponding scaling for the fluid's density, Ar for the scaled model can be found to be equal to its original value, and the ratio u_0/u_{mf} is automatically matched. Of the three scaling methodologies as well as the parcel approach discussed in this section, it is expected that this scaling methodology and law will most closely reproduce the fluidization behavior of the original scale experiment in the scaled simulation because it retains the same values for a larger set of nondimensional parameters.

In the following sections, the scaling methodologies are tested using an experimental setup and its numerical simulation at experimental scale and then scaling it to a larger scale for computationally testing of the scaling methodologies.

12.5   Description of Experimental Setup

The cold flow experiments of Sutkar et al. [5] consist of a pseudo-2D spouted fluidized bed with draft plates as shown in Fig. 12.1. The height h is the particle entrainment height below the draft plates; it is set at 0.3 m in the experiments. The draft plates address the problem of spout gas bypassing as well as the spout instability by imposing a restriction on the lateral particle flow between the spout and the annulus. By preventing the particles traveling downward in the annulus from

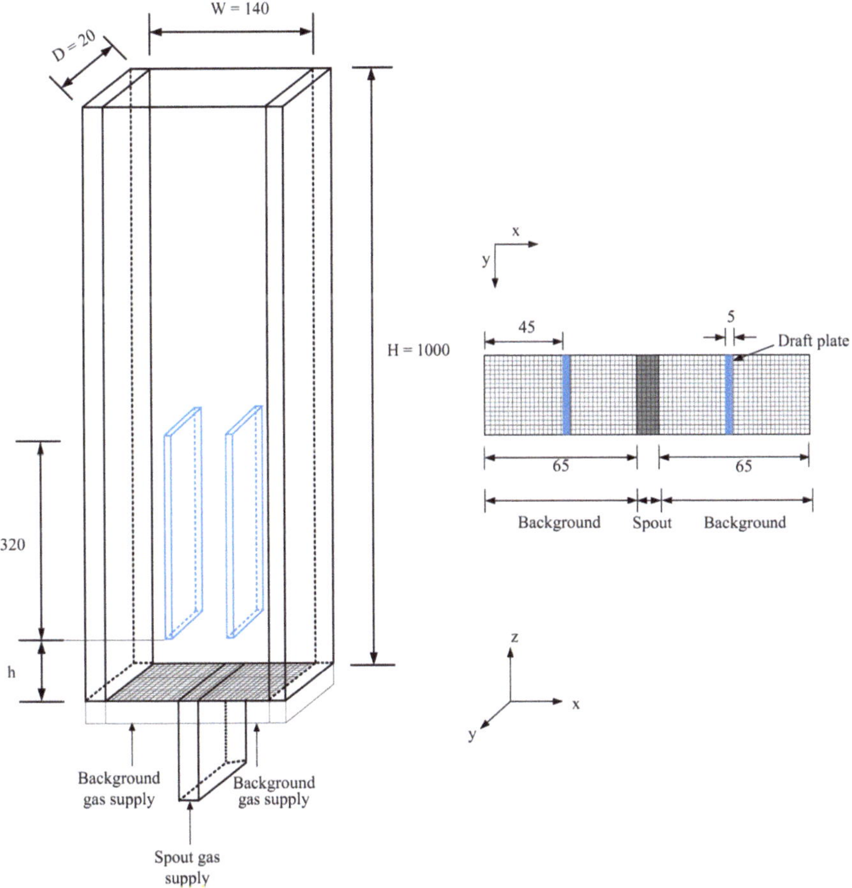

Fig. 12.1 Geometry of the spouted fluidized bed experiment of Sutkar et al. [5]

entering the spout and colliding with the particles traveling upward, the random fluctuations in the spout are eliminated [5]. In the experiments, a high-speed particle image velocimetry (PIV) camera is used to capture the instantaneous particle velocities at various heights. These velocity data are used to quantitatively compare the performance of the different scaling methodologies.

Sutkar et al. [5] employed 1-mm-diameter particles made of glass with a density of 2500 kg/m^3 or γ-Al$_2$O$_3$ particles with a density of 1040 kg/m^3 as the bed material in their experiments, although the velocity data is only available for the glass beads. The experiments consider both the "spouting with aeration" and "fluidized bed-spouting with aeration" flow regimes corresponding to different ratios of the spout velocity and background flow velocity to the minimum fluidization velocity of the particles. Here, only the fluidized bed-spouting with aeration condition is considered since it is the flow regime better suited for CLC operation. Since velocity data is not available for the γ-Al$_2$O$_3$ particles, the computational results reported by Banerjee and Agarwal [4] are also compared with a full-scale numerical study conducted as a follow-up to the experiment by the same group [9].

The simulations are performed using the commercial CFD simulation package ANSYS Fluent, release version 14.5 [1], as described in Chap. 4. The flow field is computed using the Navier-Stokes equations of fluid motion; the motion of the particles is obtained using Newton's second law. In order to achieve a coupled CFD-DEM simulation for the multiphase flow, source terms are introduced in the Navier-Stokes momentum equations to capture the solid-gas momentum exchange and in the Newtonian equation of motion to account for forces on the solid particles due to the fluid. Since simulations in this chapter are performed within a cold flow framework, the energy and species transport equations are not implemented, and the source term for interphase mass transfer is identically zero.

12.6   Computational Setup

For each bed material, glass beads, and γ-Al$_2$O$_3$ particles, four different cases are considered that reduce the total number of particles in the bed: the parcel approach, the Glicksman scaling law [2], the Link scaling law [3], and the Banerjee-Agarwal scaling law based on the terminal velocity. The independent variable for the parcel approach and the Link et al.'s scaling law are the particle scale factor n. A value of $n = 2$ is chosen for the current simulations while keeping the bed geometry the same. On the other hand, the independent variable for the Glicksman et al.'s scaling law and Banerjee-Agarwal scaling law is the geometry scale factor r; it is set at $r = 0.25$ in line with Glicksman et al. [2], and the particle scales are adjusted as determined by the respective scaling methodology. In all eight cases, u_{sp} and u_{bg} are set at roughly 37.0 u_{mf} and 1.275 u_{mf}, respectively, in accordance with the experiment to model the "fluidized bed-spouting with aeration" flow regime. Although the flow in the spouted fluidized bed setup is turbulent, it is well established that for gas-solid flows, the effect of turbulence is negligible compared to the effect of the solids for

Table 12.1 Summary of scaled test cases with glass beads with adjusted physical properties of gases and solids and comparison of nondimensional parameters with the exact scale

Case		A1	A2	A3	A4
Parameter		Exact scale	Parcel approach	Glicksman et al.'s scaling law	Link et al.'s scaling law	Banerjee & Agarwal's u_t-based scaling law
n	1	2	0.572	2	0.707
r	1	1	0.25	1	0.25
ρ_p, kg/m^3	2500	2500	2500	1251	2500
ρ_f, kg/m^3	1.225	1.225	1.225	1.225	3.463
μ_f, kg/(m-s)	1.79E-05	1.79E-05	1.79E-05	3.58E-05	1.79E-05
u_{mf}, m/s	0.66	0.66	0.33	0.66	0.33
u_t, m/s	7.75	7.75	4.62	7.75	3.87
u_{bg}, m/s	0.84	0.84	0.42	0.84	0.42
u_{sp}, m/s	24.2	24.2	12.1	24.2	12.1
$\dfrac{(u_0/u_{mf})_{sc}}{(u_0/u_{mf})_{ex}}$	1	N/A	1	1	1
$\dfrac{\mathrm{Re}_{p,sc}}{\mathrm{Re}_{p,ex}}$	1	N/A	0.286	1	1
$\dfrac{\mathrm{Ar}_{sc}}{\mathrm{Ar}_{ex}}$	1	N/A	0.187	1	1
$\dfrac{(u_0/u_t)_{sc}}{(u_0/u_t)_{ex}}$	1	N/A	0.748	0.707	1
Total # of particles	461 k	51 k	33 k	51 k	17 k
Particle time step, s	–	2e-5	5e-6	2e-5	5e-6

solid volume fractions above 0.001 [6]. Therefore, in the simulations the effect of turbulence is ignored without loss of accuracy in line with the previous work employing the CFD-DEM approach [7]. The initial bed height (equal to the bed width) is achieved by releasing a large number of particles into the bed prior to the start of each simulation. The different simulation cases and the simulation parameters are summarized in Tables 12.1 and 12.2 for glass beads and γ-Al$_2$O$_3$ particles, respectively.

The particle scaling factor n for the parcel approach in Tables 12.1 and 12.2 is the ratio between the parcel diameter and the particle diameter. The parcel approach reduces the computing cost by considering a single representative particle within each parcel to calculate the motion of the parcel. As such, a direct comparison between the nondimensional numbers does not apply for the parcel approach. It should be noted that for the 1/4 scale models, because of the reduced particle size, the particle time step is reduced from 2e-5 s to 5e-6 s to ensure that the particle collisions are accurately resolved. This increase in computational cost is more than offset by the reduced number of particles to track and is further offset by the coarser mesh required for the 1/4 scaled models to ensure that the particle volume remains smaller than the minimum cell volume. Given that the goal of the scaling methodology is to

Table 12.2 Summary of scaled test cases with γ-Al$_2$O$_3$ particles with adjusted physical properties of gases and solids and comparison of nondimensional parameters with the exact scale

Case		B1	B2	B3	B4
Parameter	Exact scale	Parcel approach	Glicksman et al.'s scaling law	Link et al.,' scaling law	Banerjee & Agarwal's u_t-based scaling law
n	1	2	0.62	2	0.708
r	1	1	0.25	1	0.25
ρ_p, kg/m^3	1040	1040	1040	520.6	1040
ρ_f, kg/m^3	1.225	1.225	1.225	1.225	3.461
μ_f, kg/(m-s)	1.79E-05	1.79E-05	1.79E-05	3.58E-05	1.79E-05
u_{mf}, m/s	0.36	0.36	0.18	0.36	0.18
u_t, m/s	4.99	4.99	2.88	4.99	2.49
u_{bg}, m/s	0.46	0.46	0.23	0.46	0.23
u_{sp}, m/s	13.2	13.2	0.66	13.2	6.6
$\dfrac{(u_0/u_{mf})_{sc}}{(u_0/u_{mf})_{ex}}$	1	N/A	1	1	1
$\dfrac{Re_{p,sc}}{Re_{p,ex}}$	1	N/A	0.310	1	1
$\dfrac{Ar_{sc}}{Ar_{ex}}$	1	N/A	0.239	1	1
$\dfrac{(u_0/u_t)_{sc}}{(u_0/u_t)_{ex}}$	1	N/A	0.711	0.707	1
Total # of particles	460 k	51 k	23 k	51 k	17 k
Particle time step, s	–	2e-5	5e-6	2e-5	5e-6

reduce the computing cost of the CFD-DEM simulation, the reduction in the total number of particles from roughly 461,000 in the original system is an important parameter for evaluating the performance of the scaling methodology. The reduction in the number of particles is the cube of the ratio of the particle scale factor n to the geometry scale factor r. For each case, a greater reduction can be achieved by increasing the independent scaling factor, but a drastic increase can alter the fluidization behavior of the system (e.g., the particles may no longer fall in the spout-able range). For the simulation cases considered, it can be seen from Tables 12.1 and 12.2 that the Banerjee and Agarwal's u_t-based scaling law offers the largest reduction in the number of particles by up to 95% leading to substantial reductions in the computing time in both cases.

The computational domain used in the simulations is an exact representation of the experimental apparatus of Sutkar et al. [4] shown in Fig. 12.1. A quarter-scale domain is used in the simulations using the Glicksman et al.'s and Banerjee and Agarwal's scaling approaches ($r = 0.25$). A structured mesh is generated with 24,000 cells for the original model and with 4000 cells for the 1/4 scale model. The difference in number is to ensure that the minimum cell volume remains greater than the particle volume for each simulation, a constraint imposed by the CFD-DEM approach. The meshes used in the simulations are shown in Fig. 12.2. Particle

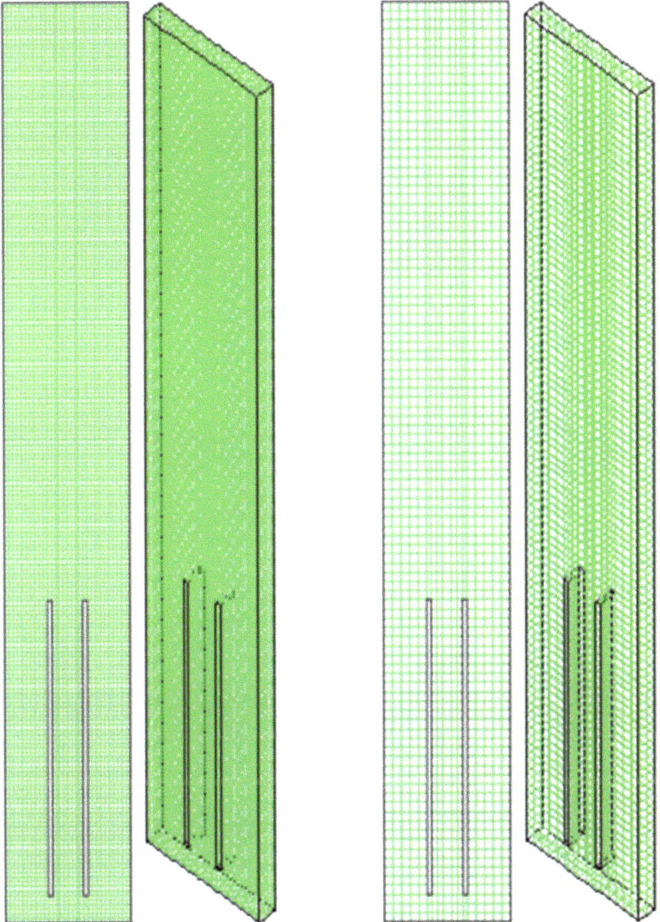

Fig. 12.2 Computational mesh for original scale model (left) and the 1/4 scale model (right)

velocity data in the z-direction is recorded in the central x-z plane at heights of 30 and 50 cm (7.5 and 12.5 cm in the 1/4 scale models) for comparison with the experiment of Sutkar et al. [4] and with the full-scale simulations conducted by the same group [9]. For simulations with the γ-Al_2O_3 particles, only the full-scale simulation results are considered for comparison since the detailed experimental data is not available.

12.7 Simulations with Glass Beads

For each simulation with glass beads as the bed material, in cases A1–A4, the particle tracks inside the spouted fluidized bed apparatus are recorded after 1 s and are qualitatively compared against the experimental data as shown in Fig. 12.3. It should be noted that while the particle tracks are instructive for verifying that the

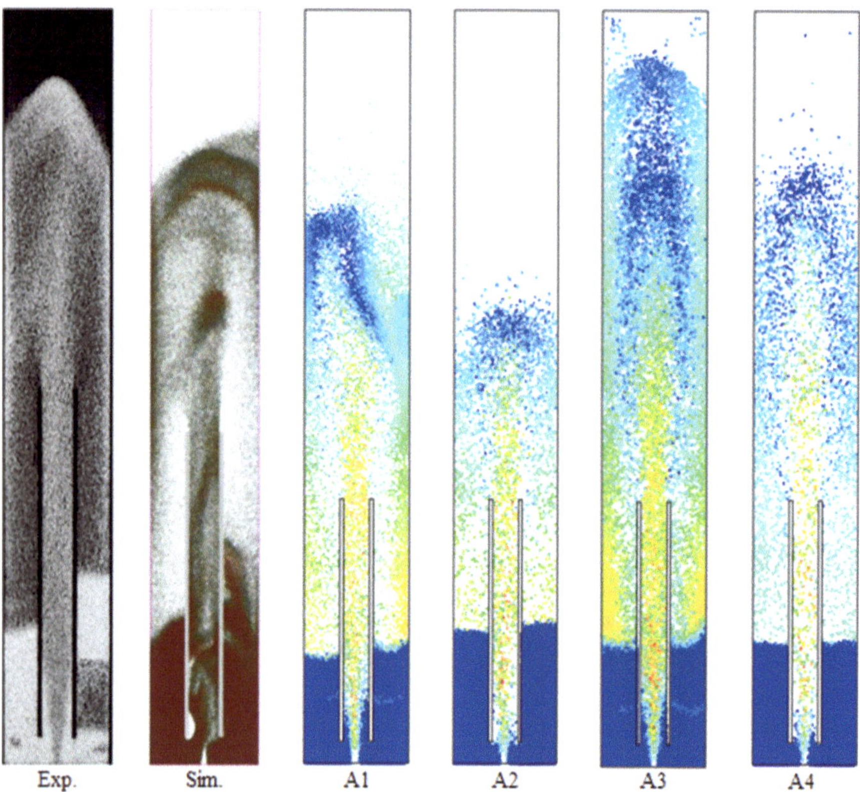

Fig. 12.3 Particle tracks after 1 s for scaled simulation cases A1–A4 compared to the experiment [4] and full-scale simulation results [9] (particles tracks are colored by the nominal velocity magnitude)

general fluidization behavior of the system remains the same in the scaled simulations, an instantaneous snapshot is not a suitable metric for determining the performance of the scaling approaches.

The 1/4 scale simulations in Fig. 12.3 (i.e., A2 and A4) are shown at the original scale for easier side-by-side comparison. Each scaled case still shows the "fluidized bed-spouting with aeration" behavior observed in the experiment despite any changes in the physical properties required by the respective scaling methodologies. The flow in the annulus is very similar for each case and matches the experimental behavior. It has been noted by Sutkar et al. [4] that the asymmetry in the annulus seen in the experiment was due to slightly uneven gas distribution and is not of any behavioral significance. The behavior of the particles observed in Fig. 12.3 is an improvement over the full-scale simulation of Sutkar et al. [9] In their simulation, the particles inside the draft plates formed clusters that partially blocked the flow leading to an unsteady pulsating flow with varying bed height. None of the present cases A1–A4 indicate the presence of particle clusters inside the draft plates, which is more in line with the experimental results than the full-scale simulation.

The spouting particles in the scaled simulations using the parcel approach and the Glicksman et al.'s scaling law do not reach the same height as the experiment, whereas the Link et al.'s scaling law slightly exceeds the experimental height; however, the Banerjee and Agarwal's u_t-based scaling law comes closest to the experimental bed height in the instantaneous snapshot. The terminal velocity is the lowest velocity required to lift a particle and carry it out of the fluidized bed. In turn, u_0/u_t is a measure of the excess energy available in the flow to lift the particles to a certain height. Since case A4 is the only scaling methodology that matches the value of u_0/u_t in the experiment, it makes sense that it is the only case that matches the bed height as well. It is surprising that the particles using the Link et al.'s scaling law reach a greater height than the experiment considering the value of u_0/u_t for case A3 is lower than that in the experiment. This can be attributed to the fact that the scaled system in A3 uses larger particles with a lower particle density than the experiment according to the scaling methodology of Link et al. [3] described in Sect. 12.3; such particles are inherently more spout-able than the original particles used in the experiment according to Geldart's powder classification [9]. It makes sense that the height of the particle tracks in case A2 has the largest discrepancy with the experiment because the scaled model according to the Glicksman et al.'s scaling law has the most unmatched nondimensional parameters compared to the experiment as shown in Table 12.1. Surprisingly, case A1 still attains a reasonable approximation to the bed height despite the relative simplicity of the parcel approach although the spout becomes asymmetric.

The particle tracks in Fig. 12.3 confirm that the general behavior of the spouted bed remains unchanged in the scaled simulations considered. However, in order to accurately characterize the performance of the various scaling methodologies, it is important to consider the particle velocities. The time-averaged particle velocity in the z-direction at two different bed heights of 30 cm and 50 cm is used to quantitatively characterize the accuracy of the different scaling methodologies. Figure 12.4 compares the time-averaged particle z-velocity in the central x-z plane at a height of 30 cm for the scaled simulation cases A1–A4 against the experimental results [4] and the results of the full-scale simulation conducted by Sutkar et al. [9].

The full-scale simulation with 460 k particles can predict the peak particle velocity in the spout region, but the predicted velocities are lower near the ends of the spout region adjacent to the draft plates. Both the Glicksman et al.'s and Link et al.'s scaling laws can capture the trend in particle velocities in the spout, but the exact velocities are lower than the experiment. This discrepancy is expected according to Table 12.1, which shows that u_0/u_t is around 0.7 for both these scaling methodologies. On the other hand, the Banerjee and Agarwal's u_t-based scaling law can accurately capture the spout velocity across the entire spout region, though it slightly overshoots the peak velocity at the center of the spout; the Banerjee and Agarwal's scaling law also shows the best results in the annulus. None of the other scaled simulations, or even the full-scale simulation, come close to matching the downward particle velocities in the annulus with the experiment. It should be noted that the parcel approach performed the worst out of all the scaling methodologies considered, which can be explained by the relative simplicity of the approach and the lack of scientific rigor in its formulation.

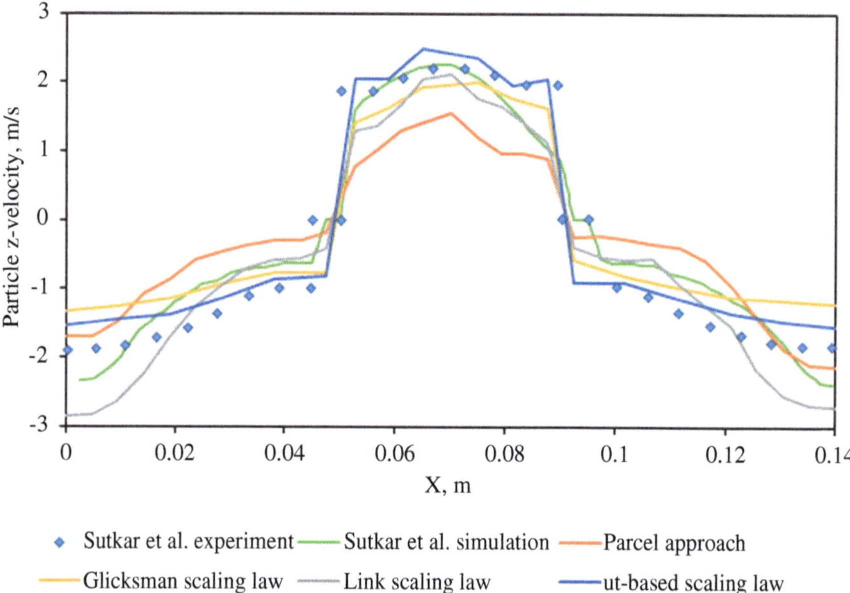

Fig. 12.4 Time-averaged particle z-velocity at $z = 30$ cm for scaled simulation cases A1–A4 compared to experiment [4] and full-scale simulation results [9]

The time-averaged particle z-velocity in the central x-z plane at a height of 50 cm for the scaled simulation cases A1–A4 is compared against the experimental results [4] and the results of the full-scale simulation conducted by Sutkar et al. [9] in Fig. 12.5. In this case, there is no clear demarcation between the spout and annulus regions because 50 cm is above the draft plates. The full-scale simulation is a good approximation of the experimental values everywhere except at the extreme ends of the bed. On the other hand, the Banerjee and Agarwal's u_t-based scaling law predicts the experimental values correctly at the extreme ends as well as in the center, but the particle velocities in other areas are too low. However, the Banerjee and Agarwal's scaling law still performs the best among the various scaling laws considered; the Glicksman et al.'s and Link et al.'s scaling laws underpredict the particle velocities across the entire bed. Once again, the parcel approach performs the worst.

12.8   Simulations with γ-Al₂O₃ Particles

The dynamic behavior and fluidization in a gas-solid system for CLC depend on the physical properties of the bed material (particle diameter, density, restitution coefficient, etc.). Therefore, it is important to verify the effectiveness of the Banerjee and Agarwal's u_t-based scaling approach for a different bed material. The γ-Al₂O₃ particles considered in this section have the same diameter as the glass beads but

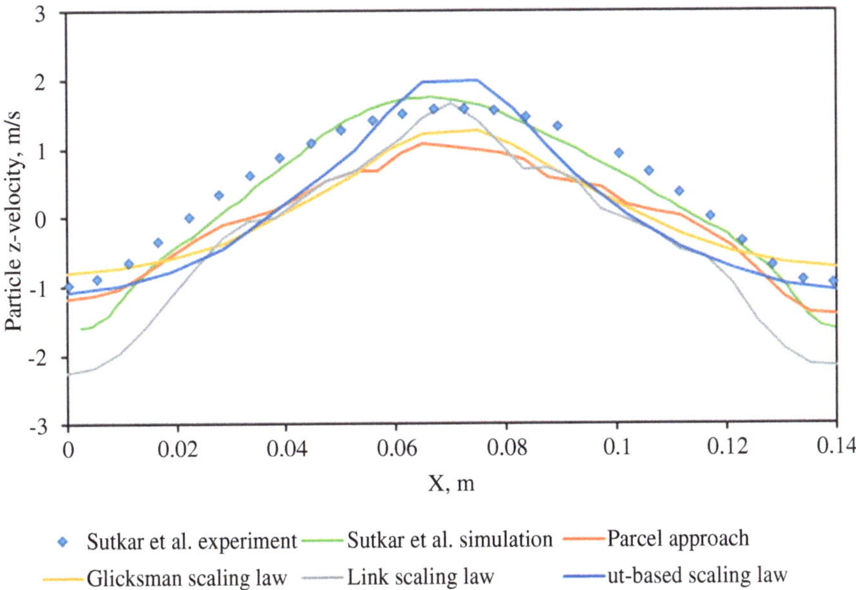

Fig. 12.5 Time-averaged particle z-velocity at $z = 50$ cm for scaled simulation cases A1–A4 compared to experiment [4] and full-scale simulation results [9]

lower density and restitution coefficient of 1040 kg/m^3 and 0.74, respectively, compared to 2500 kg/m^3 and 0.97 for glass beads. The time-averaged particle velocity of the γ-Al$_2$O$_3$ particles in the z-direction for simulation cases B1–B4 in Table 12.2 is recorded at heights of 10 and 30 cm, respectively, and compared against the results from the full-scale simulation of Sutkar et al. [9] The velocity profiles using the different scaling methodologies listed in Table 12.2 at 10 cm and 30 cm are presented in Figs. 12.6 and 12.7, respectively.

It can be noted that the full-scale simulation results shown in Figs. 12.6 and 12.7 are asymmetric due to the way the boundary conditions were imposed [9]; similar asymmetry could be observed in the particle tracks shown in Fig. 12.3 as well as in the velocity profiles in Figs. 12.4 and 12.5. As such, only the magnitude of the velocities in the full-scale simulation results should be considered for comparing the scaling methodologies, not the profile shape. At $z = 10$ cm, there is a strong spout in the central region, but the particles in the annulus are densely packed. On the other hand, at $z = 30$ cm, the spout velocities are weaker, but there is a distinct downward motion of the particles in the annulus. In both Figs. 12.6 and 12.7, the Link et al.'s scaling law and Banerjee and Agarwal's u_t-based scaling law provide the best match with the full-scale simulation results; the parcel approach and the Glicksman et al.'s scaling law underpredict the spout velocity. All the scaling laws capture the densely packed bed with no particle velocity in the annulus as shown in Fig. 12.6. However, the Banerjee and Agarwal's u_t-based scaling law better captures the gradual change in particle z-velocities in the annular region as shown in Fig. 12.7; the Link et al.'s

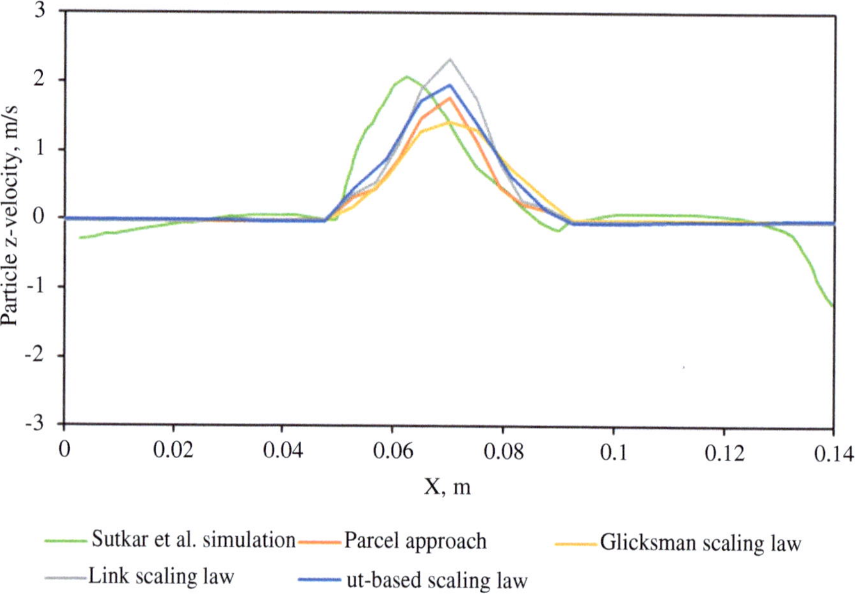

Fig. 12.6 Time-averaged particle z-velocity at $z = 10$ cm for scaled simulation cases B1–B4 compared to full-scale simulation results [8]

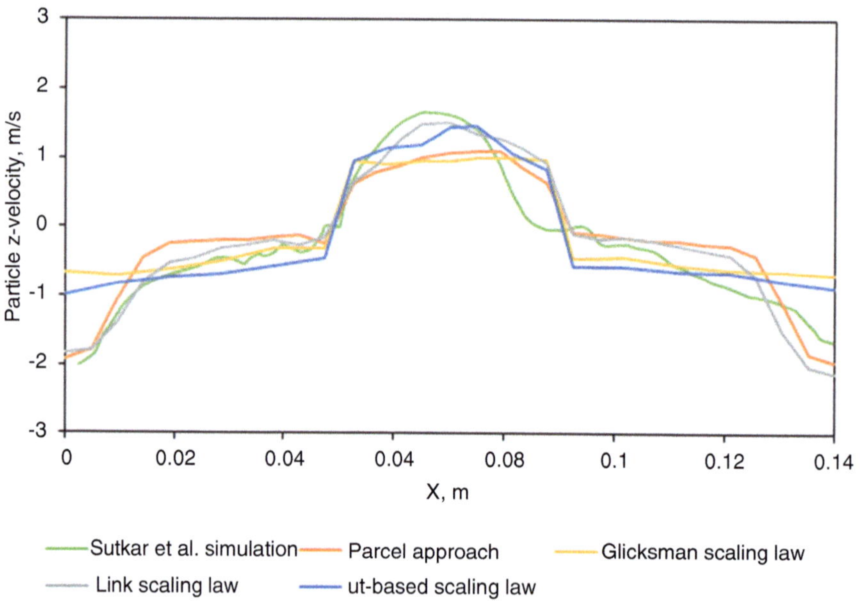

Fig. 12.7 Time-averaged particle z-velocity at $z = 30$ cm for scaled simulation cases B1–B4 compared to full-scale simulation results [9]

scaling law shows a nearly zero velocity in much of the annulus before producing a sharp decrease toward the edges. Once again, the Banerjee and Agarwal's scaling law can be seen to produce the best match with the full-scale simulation results on top of providing the largest reduction in the number of particles.

Based on the results presented in this chapter, it can be concluded that Banerjee and Agarwal's terminal velocity-based scaling laws capture the performance of a spouted fluidized bed at different scales more accurately compared to other scaling laws reported in the literature.

References

1. A.F.U. Guide, *Ansys Fluent Theory Guide* (ANSYS Inc., Novemb, 2013)
2. L.R. Glicksman, M. Hyre, K. Woloshun, Simplified scaling relationships for fluidized beds. Powder Technol. **77**(2), 177–199 (1993)
3. J.M. Link, W. Godlieb, P. Tripp, N.G. Deen, S. Heinrich, J.A.M. Kuipers, M. Schönherr, M. Peglow, Comparison of fibre optical measurements and discrete element simulations for the study of granulation in a spout fluidized bed. Powder Technol. **189**(2), 202–217 (2009)
4. S. Banerjee, R.K. Agarwal, Characterization of scaling Laws in computational fluid dynamics simulations of spouted fluidized beds for chemical looping combustion. Energy Fuel **30**(10), 8638–8647 (2016)
5. V.S. Sutkar, T.J.K. van Hunsel, N.G. Deen, V. Salikov, S. Antonyuk, S. Heinrich, J.A.M. Kuipers, Experimental investigations of a pseudo-2D spout fluidized bed with draft plates. Chem. Eng. Sci. **102**, 524–543 (2013)
6. J.K. Claflin, A.G. Fane, Spouting with a porous draft-tube. Can. J. Chem. Eng. **61**(3), 356–363 (1983)
7. S. Elghobashi, On predicting particle-laden turbulent flows. Appl. Sci. Res. **52**(4), 309–329 (1994)
8. J.M. Parker, CFD model for the simulation of chemical looping combustion. Powder Technol. **265**, 47–53 (2014)
9. V.S. Sutkar, N.G. Deen, B. Mohan, V. Salikov, S. Antonyuk, S. Heinrich, J.A.M. Kuipers, Numerical investigations of a pseudo-2D spout fluidized bed with draft plates using a scaled discrete particle model. Chem. Eng. Sci. **104**, 790–807 (2013)

Chapter 13
Machine Learning for Chemical Looping Combustion

Mitchell [1] defines ML as "A computer program is said to learn from experience E with respect to some class of tasks T and performance measure P, if its performance at tasks in T, as measured by P, improves with experience E." Domingos et al. [2] define that machine "learning = representation + evaluation + optimization." "Representation" aims to describe the model with the language that the computer can interpret. The represented models will then be scored by "evaluation" and "optimization" to search for the high-scoring one. ML is classified into supervised learning, unsupervised learning, semi-supervised learning, reinforcement learning, etc. Supervised learning requires dataset with labeled samples, unsupervised learning does not need labeled inputs, semi-supervised learning needs a small number of labeled samples, and reinforcement learning requires a feedback mechanism. In this chapter, the concepts of different ML methods and typical algorithms are introduced, which are followed by two typical applications of ML to chemical looping.

13.1 Machine Learning Fundamentals

13.1.1 Supervised Learning

Supervised learning [3] is a powerful tool to classify and process data using ML. The "supervised" means the learning process is accomplished under the seen label of observation variables. In supervised learning, the datasets are classified into the training set and the testing set. The training set is used to build the ML model, where the input variables are the features which influence the accuracy of the predicted variable including the quantitative and qualitative variables, and the output variable is the label class that supervised learning labels as new observations. The supervised learning task can be classified into two types, namely, the classification task where the output variables are categorical variables and the regression task where the output variables are continuous variables.

R. K. Agarwal, Y. Shao, *Modeling and Simulation of Fluidized Bed Reactors for Chemical Looping Combustion*, https://doi.org/10.1007/978-3-031-11335-2_13

In the supervised learning process, $\{(x^{(i)}, y^{(i)}), i = 1, 2, \ldots, m\}$ is the training set, where $x^{(i)}$ is the input variable, $y^{(i)}$ is the output variable, and a pair $(x^{(i)}, y^{(i)})$ is a training example. In general, X and Y denote the space of input and output variables, respectively. The training procedure is aimed at obtaining a function $h\colon X \to Y$. Thus, for an input variable x in the testing set, the corresponding value of y can be obtained via the predictor $h(x)$. In this section, five commonly used supervised learning methods are introduced including logistic regression, decision tree, random forest, neural network, and support vector machine.

13.1.1.1   Logistic Regression

Logistic regression is one of the simplest ML classification algorithms in supervised learning. It is one of the generalized linear models and is commonly applied to binary classification of the samples. The sigmoid function which maps a predicted real value to a probabilistic value between 0 and 1 is used as a cost function [4]. The binary logistic regression model is expressed as [5]

$$p_i = f(y|x) = \frac{1}{1 + \exp\{-(\beta_0 + \beta_1 x_i)\}} \tag{13.1}$$

where p_i is the probability of occurrence of the given event, and $P(y_i = 1) = p_i$ and $P(y_i = 0) = 1 - p_i$, $0 \le p_i \le 1$. β_0 and β_1 are the model parameters and x_i is the input variable.

Logistic regression presents the advantage of high flexibility in its assumptions and wide applicability. Besides binary classification, the algorithm can also be extended to continuous variables, and the multiple logistic regression mode is given as [5]

$$p_i = \frac{1}{1 + \exp\{-z_i\}} = \frac{\exp\{z_i\}}{1 + \exp\{z_i\}} \tag{13.2}$$

where $z_i = \beta_0 + \beta_1 x_1 + \beta_2 x_2 + \ldots + \beta_n x_n$.

13.1.1.2   Decision Tree

Logistic regression is not appropriate to be applied to the problems with complex decision boundaries since it is dependent on the assumption of linear separability of feature space. Decision tree can overcome this issue by recursively partitioning the feature space into hypercuboids [6]. The structure of a decision tree is shown in Fig. 13.1, which includes a root, nodes, edges, and leaf nodes. The root is a unique node where the first split takes place; the node is the position from where the tree will split according to the independent variable and its value in the dataset; the edge

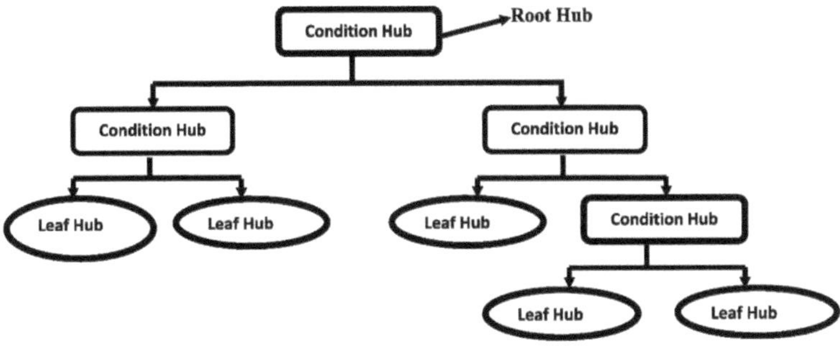

Fig. 13.1 Topological structure of decision tree [7]

connects the nodes and presents the direction of a split; the leaf node also named as terminal node is the output of the decision tree. The running of the algorithm starts from the root node and then cascades down the tree to satisfy the parameters and decision. The process continues until a terminal node is encountered and the output can be obtained.

13.1.1.3   Random Forests

Random forest is a supervised learning algorithm proposed by Breiman [8] in 2001 by using many decision trees, which can be used for classification and regression. As an ensemble learning method, random forest regression is very flexible and robust. In the training process, a large set of decision trees is created, and the results obtained from individual trees are averaged to get the final output. This technique uses bootstrap aggregation—that is each tree is trained on a random subset of the dataset and only uses some random set of the features. This procedure ensures the independent run of each tree, and the final outputs is obtained without preference to any tree. The random forest regression model has the capability to evaluate the influence of the input variables on the output.

13.1.1.4   Neural Networks

Neural network is the most well-known supervised learning method. According to the universal approximation theorem, any function can be approximated by a sufficiently large and deep network. The feedforward neural network also called multilayer perceptron (MLP) or fully connected neural network is one of the most popular neural network models. It contains a modular structure based on individual logistic regression units termed neurons as central building elements, which contributes to its great power and flexibility [9] A typical feedforward neural network model includes three orderly basic layers which are input, hidden, and output layers

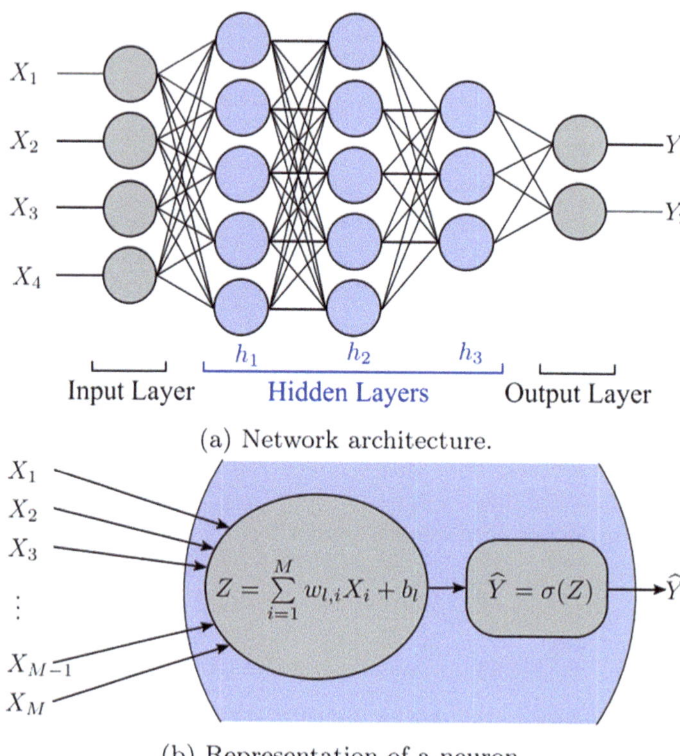

(a) Network architecture.

(b) Representation of a neuron.

Fig. 13.2 Topological structure of feedforward neural network [6]

[10]. Figure 13.2 displays the topological structure of the feedforward neural network. The input layer is where the data enters the network; the hidden layer processes the data from the first layer and then sends it to the output layer. For each neuron, it processes the input data through an activation function σ. The output of the neuron l is expressed as $\widehat{y} = \sigma(Z)$ with $Z = \sum_{i=1}^{M} w_{l,i} X_i + b_l$, where X_i is the input, $w_{l,\ i}$ is the weight of connection, b_l is the bias, Z is the input of the hidden layer neuron, and $\widehat{y}$ is the output of the hidden layer neuron. The neurons can be combined in different structures to describe the problem and the dataset. To improve the training accuracy, nonlinear optimization methods such as backpropagation can be used to adjust the network weights.

13.1.1.5　Support Vector Machines

Support vector machines were first described in the book by Vapnik [11] in 1995 and later in the book by Cristianini and Shawe-Taylor [12] The algorithm was first proposed to deal with classification tasks. Then, it was also developed to solve

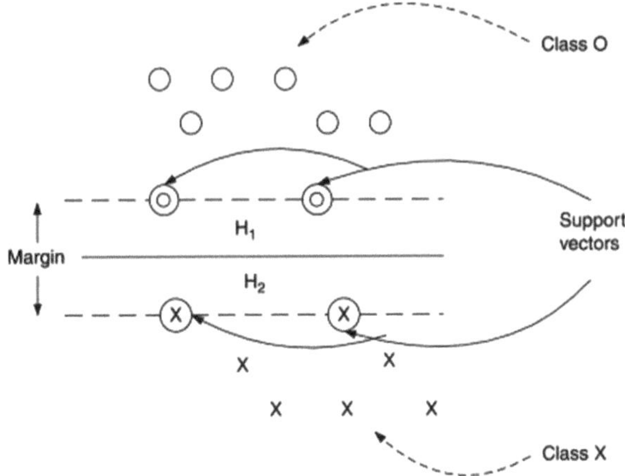

Fig. 13.3 An illustration of support vector machine classification scheme [13]

nonlinear regression problem by introducing ε-insensitive loss function. Thus, SVMs can be used for both classification and regression analysis and have the capability to handle multiple continuous and categorical variables. As shown in Fig. 13.3, in the classification analysis, the algorithm separates two classes of samples—the Class O and Class X in a multidimensional space using the hyperplane H_1 and H_2. The samples located on the boundaries of the two classes are regarded as support vectors. The goal of support vector machine classification is to maximize the margin between the support vectors using kernel function. Support vector machines are memory-efficient and versatile due to the advantage of dealing with the high-dimensional spaces.

13.1.2   Unsupervised Learning

Unsupervised learning uses algorithms for problems with unlabeled data, where the learning task is to extract features from the dataset by specifying certain global criteria without supervision [9]. Unsupervised learning is mainly used for clustering and dimensional reduction problems. Clustering is to assign similar data to different groups without any prior knowledge, which includes two main classes, namely, the hierarchical clustering and partitional clustering. Dimensional reduction technique is to identify important features from large-scale dataset and extract lower-dimensional representations for high-dimensional data that can be used for supervised learning algorithms. *k*-means clustering and principal component analysis are two most popular unsupervised learning algorithms.

Fig. 13.4 Principle of a k-means clustering algorithm [14]: (**a**) selection of initial cluster centers, (**b**) classification of points and calculation of new cluster centers, and (**c**) reclassification of the points with the new cluster centers

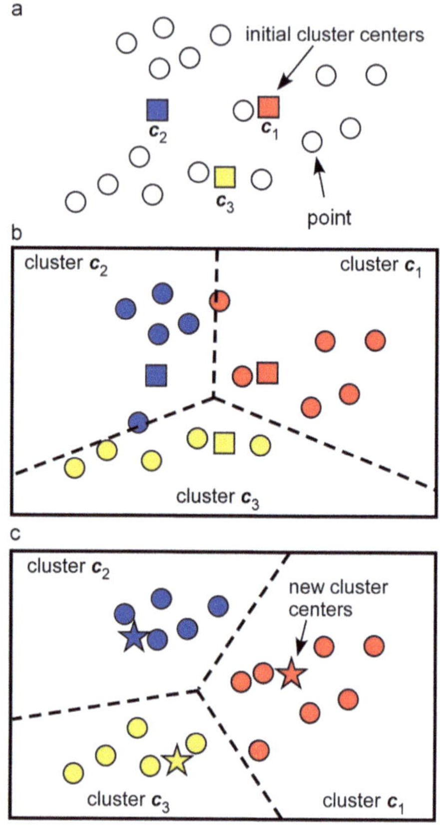

13.1.2.1 k-**Means Clustering**

k-means clustering includes two steps, a formation of reference pattern and a pattern matching process. Figure 13.4 illustrates the principle of a k-means clustering algorithm. The algorithm randomly subdivides a dataset of N points into k clusters, and the distance between each point and each cluster center is calculated. The point is classified into the cluster which contains the nearest cluster center. After the classification of all points, new cluster centers are created by averaging the points in each cluster. The process of classification and creation of new cluster centers is repeated until the realization of the convergence of a criterion function.

13.1.2.2 **Principal Component Analysis**

Principal component analysis is a widely known dimension-reduction algorithm. Principal component analysis captures a set of principal components of the redundant information in a high-dimensional correlated space and maps the data to a lower-dimensional subspace. Thus, an n-dimensional dataset can be reduced to fewer dimensions with minimal loss of information. The captured principal components are orthogonal to each other, and each describes unique information, which means the selected principal components are sufficient to reveal the main characteristics of the dataset. Principal component analysis algorithm process can be explained as [6] a matrix describing the dataset is first transformed into a symmetric covariance matrix, which is then decomposed into a matrix containing the eigenvalues and a matrix containing the principal components; and then the principal component score matrix is finally obtained by multiplying the original dataset with the eigenvector matrix. In the principal component analysis algorithm, the dimensional reduction makes the trends and correlations which are hidden in the redundant dataset more visible.

13.1.3   Semi-supervised Learning

Semi-supervised learning is used for the dataset that contains labeled and unlabeled information at the same time, and thus the algorithms operate under partial supervision.

13.1.3.1   Reinforcement Learning

Reinforcement learning has an agent and optimizes the interactions between the agent and the environment over time. When the agent interacts with the environment, it will get a reward and discover which action gets the most reward by trial and error. In the learning process, there is not labeled information about the correct actions for the agent, and the agent learns from its own experience in the form of rewards. The learning goal is to maximize the cumulative reward signal. Reinforcement learning has two important elements, the agent's policy and the value function [9] The policy implies the behavior of an agent, which is a mapping $a = \pi(s)$ between the states s of the system and the optimal action a; through the policy π, the actions of the agent in different conditions can be decided. The value function $V(s)$ refers to the utility of reaching the state s for maximizing the long-term rewards of the agent. When the instantaneous reward signal is maximized, the optimal policy can be obtained. Reinforcement learning algorithms are generally classified into two categories which are the value-based algorithm and policy-based algorithm. As repetitive account for the interaction between the agent and the environment is needed; reinforcement learning requires significant computational resources.

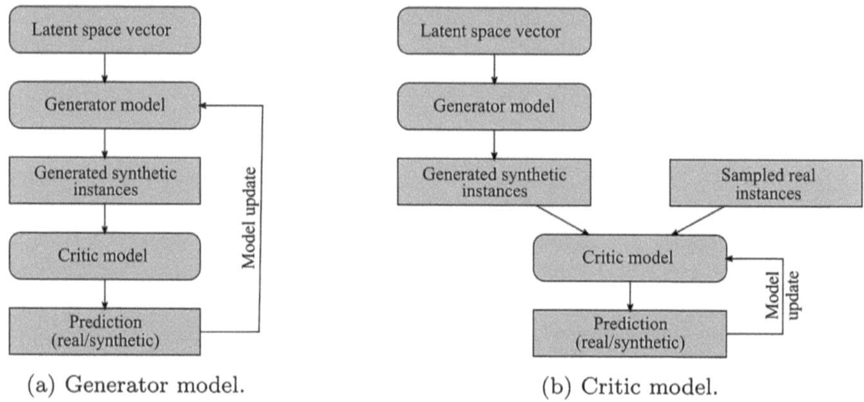

(a) Generator model.   (b) Critic model.

Fig. 13.5 The training loop of generative adversarial network algorithms: (**a**) a generator model and (**b**) a critic model [6]

13.1.3.2 Generative Adversarial Network

The generative adversarial network algorithms contain two sub-models, a generator and a critic also called discriminator. A training loop of the method is illustrated in Fig. 13.5. The generator takes the inputs from a latent space with the intention of mimicking the real target distribution and outputs synthetic data to the discriminator; the discriminator tries to distinguish the difference between the generated fake sample and the original real data through classification. The verdict of the discriminator serves as the feedback to the generator which tries to produce more realistic fake samples. The discriminator aims to maximize the probability of real data and the generated synthetic data, while the generator intends to minimize the probability. The continuous adversarial training of generator and discriminator is performed until the discriminator cannot distinguish the generated high-quality fake samples from the real samples.

13.2 Applications of Machine Learning to CLC

13.2.1 Application to Circulating Fluidized Bed Riser [15]

Circulating fluidized bed riser has the advantages of favorable gas-solid contact and reaction efficiency as well as flexible adjustment of operating conditions. Most existing CLC systems use a riser as the fuel reactor or air reactor, and even both the reactors are designed as risers in some dual circulating fluidized bed CLC systems. However, a first-principle understanding of the intense gas-solid flow regimes is still not adequate because of complexities of the fluidization phenomena in the riser. The emergence of machine learning opens up possibilities to understand the flow regimes without the requirement of detailed equations or assumptions describing the fundamental phenomena.

Chew and Cocco [15] used two machine learning methods—the random forest and neural network—to investigate the flow characteristics in a circulating fluidized bed riser. Two goals of the work are to (i) determine the relative importance of the process variables to the fluidization phenomena via random forest and (ii) establish a model with high prediction accuracy in the absence of first-principle information to understand the complex phenomena via neural network.

In a circulating fluidized bed riser, the local mass flux could be influenced by the parameters including mean particle diameter (d_{ave}), particle density (ρ_s), radial position (r/R), height along the riser (h/H), superficial gas velocity (U_s), and overall mass flux (G_s). To rank the variables in order of importance with respect to the influence on the local mass flux, random forest algorithms are performed based on a dataset with 1320 groups of data obtained from experiments. The dataset contains 660 groups of data for narrow particle size distribution, 440 groups of data for binary mixture, and 220 groups of data for broad particle size distribution system. The predicted importance estimates for local mass flux are given in Fig. 13.6. It can be seen that the riser positions including radial position and height along the riser are the two most important factors, which are followed by particle properties and operating conditions including the mean particle diameter, particle density, and superficial gas velocity, while the overall mass flux has least influence in the local mass flux. Notably, the random forest method provides the means to precisely predict the importance of each variable without the need of understanding the complex physical characteristics of the intense flow in the riser.

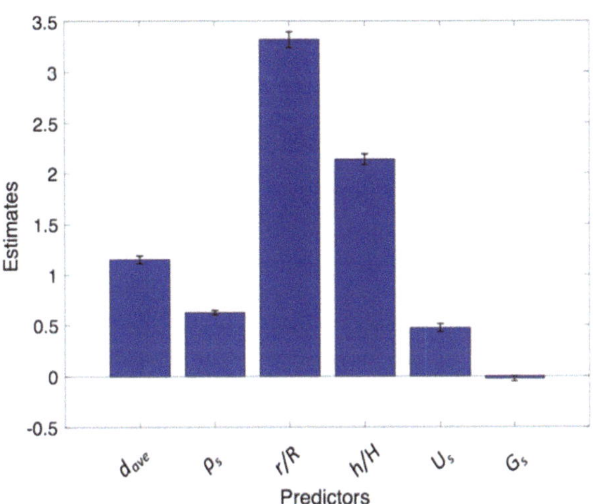

Fig. 13.6 Random forest out-of-bag permuted predictor importance estimates for local mass flux [15]

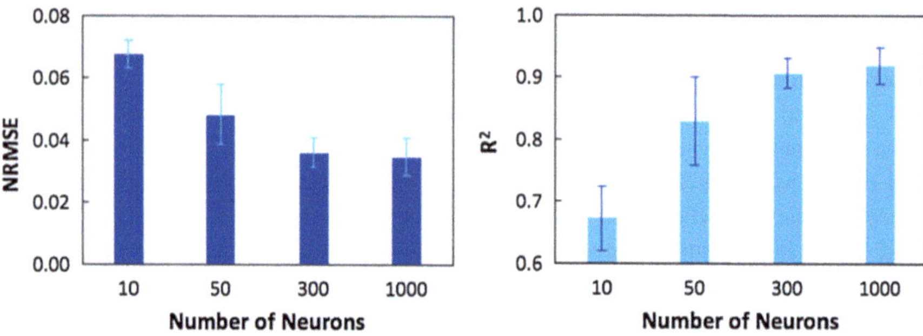

Fig. 13.7 Effects of neuron numbers of neural network on normalized root mean-squared error and coefficient of determination [15]

To establish a high-accuracy prediction model, the neural network is used with the abovementioned six parameters as the inputs (i.e., the mean particle diameter, particle density, radial position, height along the riser, superficial gas velocity, and overall mass flux) and one targeted parameter (i.e., local mass flux, local concentration, local segregation, or local cluster characteristics) as the output. The Levenberg-Marquardt backpropagation training algorithm in MATLAB is used. Two parameters, namely, the normalized root mean squared error (NRMSE) and coefficient of determination (R^2), are used to evaluate the accuracy of the prediction model. NRMSE represents the square root of the average squared difference between prediction and target, normalized with respect to the difference between the maximum and minimum; R^2 denotes the correlation between prediction and target. Figure 13.7 illustrates the effects of the number of neurons in the hidden layer on the prediction accuracy. 300 neurons are finally chosen for neural network since further increase in the neuron number has little influence on the prediction accuracy. Thus, based on the established neural network, the local mass flux can be predicted with the known six inputs avoiding the laborious and time-consuming experiments or numerical simulations.

With similar process, the relations between the six input parameters and other fluidization features such as local concentration, local segregation, and local cluster characteristics can also be predicted, which provide deep insight into the not fully understood fluidization phenomena.

13.2.2 Application to the Design of New Oxygen Carriers [16]

Oxygen carrier plays an important role in the CLC process, since it continuously transfers oxygen from the air reactor to the fuel reactor. The inability to select an optimal oxygen carrier with the consideration of the cost, availability, reactivity, stability, and oxygen transfer capacity is a key barrier in the development of CLC technology. Natural ores and some other industrial products have gained increasing

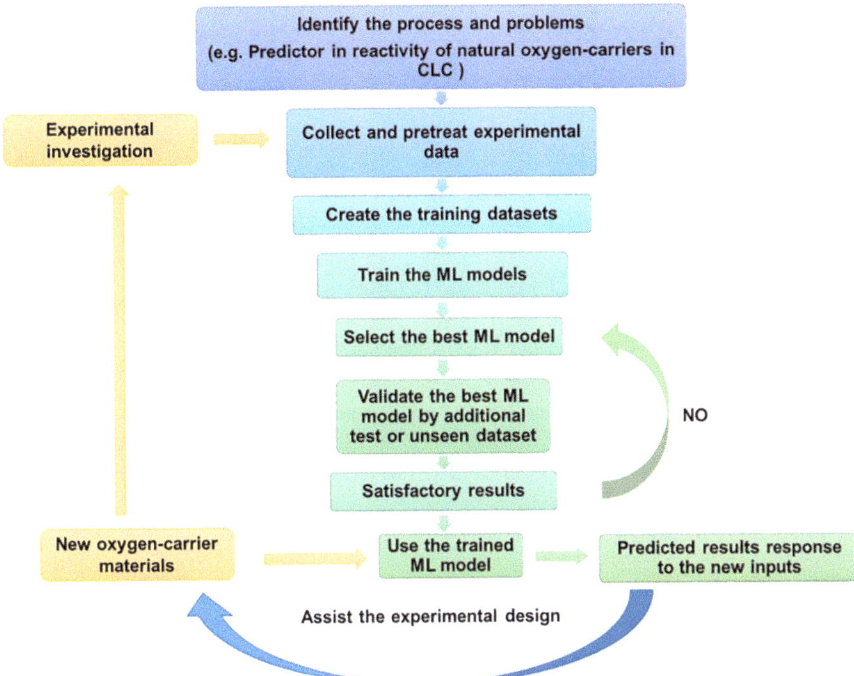

Fig. 13.8 Workflow chart for application of machine learning to the design of oxygen carriers in CLC [16]

attention because of the low cost and abundant reserves. However, their reaction performance varies significantly due to variable compositions and structures. Therefore, it is essential to investigate the heterogeneous and multicomponent materials for oxygen carriers for their performance before proceeding to large-scale industrial applications. Based on the experimental data, neural network approach was first used by Yan et al. [16] to exploit the valuable information obtained from the reaction performance of different oxygen carriers in order to assist in the design and manufacture of best/optimal oxygen carrier.

Figure 13.8 shows the workflow chart for application of machine learning to the design of optimal oxygen carriers in CLC. First, experiments were conducted to study the reaction performance of oxygen carriers in a bench scale, batch fluidized bed reactor. Thus, a dataset consisting of physical and chemical properties, experimental conditions, reactivity with fuels, and oxygen release rates was obtained consisting of 171 groups of data. The next step was to design and train a neural network model. Fourteen parameters, including the mass concentration of the elements Mn, Fe, Si, Al, Ti, Ca, K, Mg, Na, and P and bed temperature, attrition index, crushing strength, and BET surface area of the fresh material were used as the inputs; the reactivity with syngas, methane, and the oxygen release to the gas phase

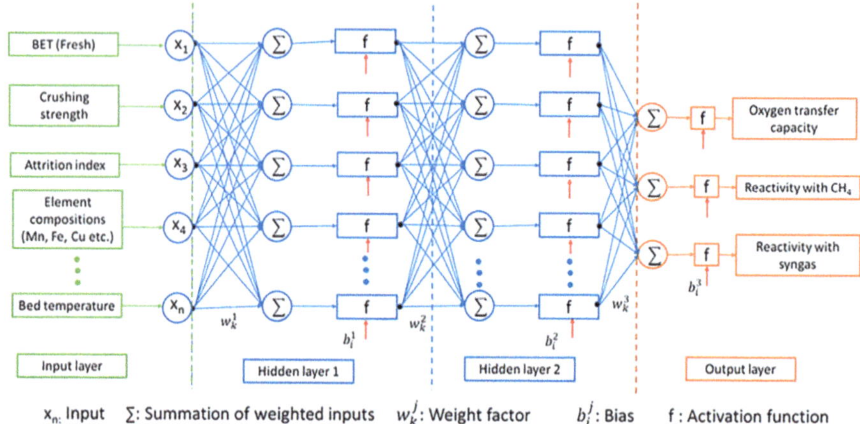

x_n: Input Σ: Summation of weighted inputs w_k^j: Weight factor b_i^j: Bias   f : Activation function

Fig. 13.9 Proposed neural networks structure to predict the reactivity of manganese ores in the chemical looping process [16]

was used as the output. Three training functions including trainlm, trainbr, and trainbfg were selected to train the multilayer neural networks shown in Fig. 13.9 in MATLAB. The number of neurons and hidden layers were decided based on the minimum value of mean absolute error (MAE) and the highest value of the coefficient of determination (R^2). Afterward, to validate the accuracy and robustness of the established model, unseen data that was not included within the training dataset was used for the model. The predicted results response to the new inputs provided valuable guidance for the design of oxygen carriers.

13.2.3   Application to Optimization of a Moving Bed Reactor [17]

A tower-type moving bed can be used as the air reactor in a chemical looping combustion system because of its low-pressure drop and smooth operation. A quasi-two-dimensional numerical model was established using DEM approach to investigate the velocity and solid residence time distributions in the moving bed [18] as shown in Fig. 13.10. Further, the flow patterns under different operating and structural parameters can be optimized via machine learning methods. Random forest regression model is applied to evaluate the importance of each variable to the solid flow pattern, while the feedforward neural network is applied to build up a high-accuracy model to predict the solid axial velocity in the moving bed without the requirement to understand the physical mechanisms.

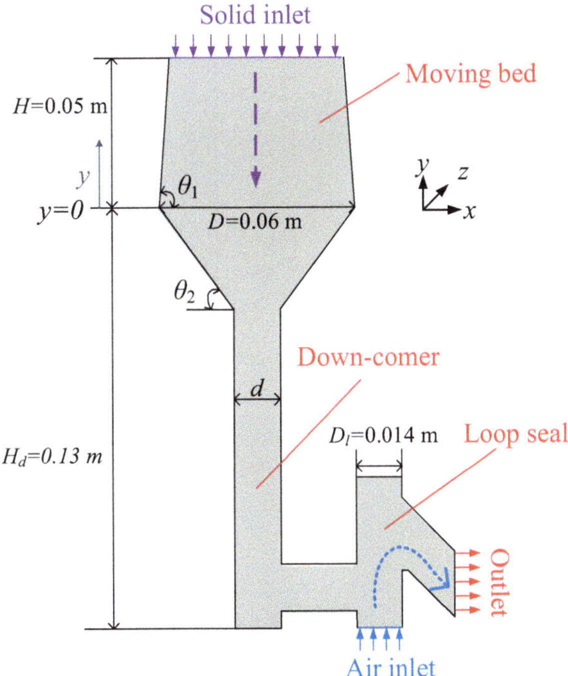

Fig. 13.10 Schematic of moving bed air reactor [18]

Table 13.1 Details of the database obtained from the CFD-DEM simulation

Reactor angle θ_1	Wedge angle θ_2	Diameter ratio of downcomer to reactor d/D	Solid mass flux J_p (kg/m^{2}s)	Mass flow index MFI	Solid axial velocity (m/s)
75–85°	40–55°	0.17–0.27	50–150	0.004–0.911	0.01–4.98

In Fig. 13.10, particles have a density of 2600 kg/m^3 and a diameter of 0.45 mm. The sphericity is set to 1. Air temperature is set to 20 °C. Air is introduced to the loop seal, which transports the particles to the exit. To remain the constant number of particles in the moving bed, particles are fed from the top inlet with the flow rate same as that in the outlet. Thus, continuous solid flow is realized with the steady flow rate. The solid mass flux can be adjusted via control of inlet gas velocity in the loop seal. More detailed description of the CFD-DEM simulation can be found in literature [18]. The dataset is composed of 1911 groups of velocity data which are measured at different radial and axial positions of the moving bed at different solid mass fluxes J_p, reactor angles θ_1, wedge angles θ_2, and ratio of downcomer diameter to reactor diameter d/D. Table 13.1 gives details of the database obtained from the CFD-DEM simulation.

Table 13.2 Influence of tree number on the errors and coefficient of determination

Tree number	RMSE	MAE	R^2
300	0.020	0.012	0.988
400	0.021	0.014	0.989
500	0.023	0.014	0.987
600	0.020	0.013	0.989
700	0.024	0.014	0.985
800	0.021	0.014	0.988
900	0.022	0.013	0.986
1000	0.019	0.014	0.988

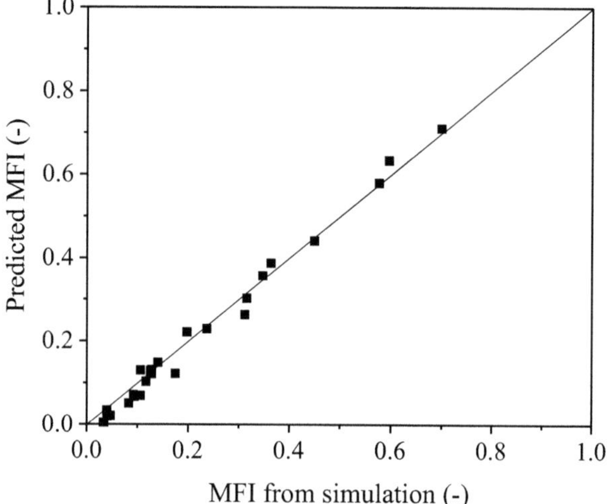

Fig. 13.11 MFI predicted from the random forest model vs. MFI obtained from the numerical simulation [17]

The random forest regression method is adopted to evaluate the importance of each variable to mass flow index (MFI) which is the average particle velocity ratio between the near-wall zone and the central zone: MFI $= V_{\text{wall}}/V_{\text{centerline}}$. The database is randomly separated into two distinct subgroups, namely, training set (90% of the databank) and test set (10% of the databank). The inputs are the reactor angle θ_1, wedge angle θ_2, and ratio of downcomer diameter to reactor diameter d/D, bed height h/H, and solid mass flow rate J_p; the output is mass flow index. The number of trees may affect the prediction accuracy. Table 13.2 illustrates effects of tree number on the errors and coefficient of determination. When the tree number is larger than 300, the errors and coefficient of determination don't change much. Thus, 300 trees are finally determined for the random forest regression model in this application. The values of coefficient of determination R^2 between the predictions of the random forest regression model and the simulation results are close to 1. This demonstrates good agreement between the predictions and simulation data, which can also be seen in Fig. 13.11. As a measure of accuracy, the values of root-mean-

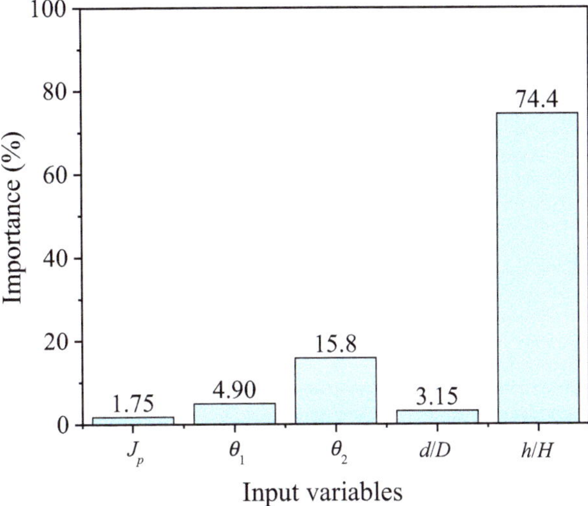

Fig. 13.12 Random forest out-of-bag permuted predictor importance estimates for the mass flow index [17]

square error (RMSE) and mean absolute error (MAE) indicate the excellent performance of the established random forest regression model for predicting the mass flow index in the moving bed.

Random forest model has the capability to perform sensitivity analysis to understand the contribution of each input variable to the accuracy of the output. Figure 13.12 illustrates the relative importance of each independent parameter for the mass flow index. The height along the reactor h/H has the highest impact on the mass flow index, followed by wedge angle θ_2, reactor angle θ_1, and ratio of downcomer diameter to reactor diameter d/D; the solid mass flux J_p has the minimum effect. The results agree with those obtained from Pearson correlation test.

A feedforward neural network model then is developed by adopting solid axial velocity as the output and using the reactor angle θ_1, wedge angle θ_2, and the ratio of downcomer diameter to reactor diameter d/D, radial position r/R, and axial position y/H as the inputs, to predict the solid flow velocity in the moving bed. The solid mass flux is not set as the input because of its very little influence on the mass flow index, which has been demonstrated in the last section. All the data is normalized via the map minmax function. The data are trained in the network among which 90% is used as the training set, and 10% is used as the testing set. The Levenberg-Marquardt algorithm is used for the data training since it is the fastest training algorithm for small- and medium-scale feedforward network. The training number is set to 5000, and the network learning precision is set to 0.00001.

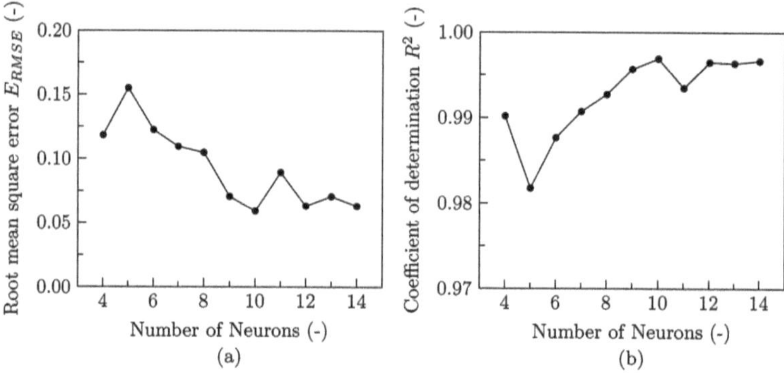

Fig. 13.13 Effects of number of the neurons on the prediction accuracy of the BP neural network model [17]: (**a**) root mean square error distribution and (**b**) coefficient of determination distribution

Figure 13.13 displays the effects of neuron number on prediction accuracy of the neural network model. When the neural number is increased from 4 to 14, the root mean square error E_{RMSE} is smallest of 0.059, and the coefficient of determination R^2 is largest of 0.997 at the neuron number of 10. Thus, ten neurons are finally determined for the feedforward neural network.

The established feedforward neural network mode is further used for prediction of the axial velocity distributions under different geometric parameters, and the corresponding mass flow indices and effective transition point (ETP) heights h_c/H, which is the interface between the mass flow and the funnel flow, are calculated via preorder traversal. Figure 13.14 displays the relation between the ETP height h_c/H and the structural parameters. To design a moving bed reactor with an expected h_c/H, there are different geometry designs which can meet the requirement. If the least half of the reactor should be designed to operate in the mass flow pattern, the reactor geometry with $\theta_1 = 80 - 85°$, $\theta_2 = 55°$, and $d/D = 0.18 - 0.20$ can be selected. The successful establishment of the BP neural network model provides valuable guidance for the reactor design and optimization without complex simulation and experiments.

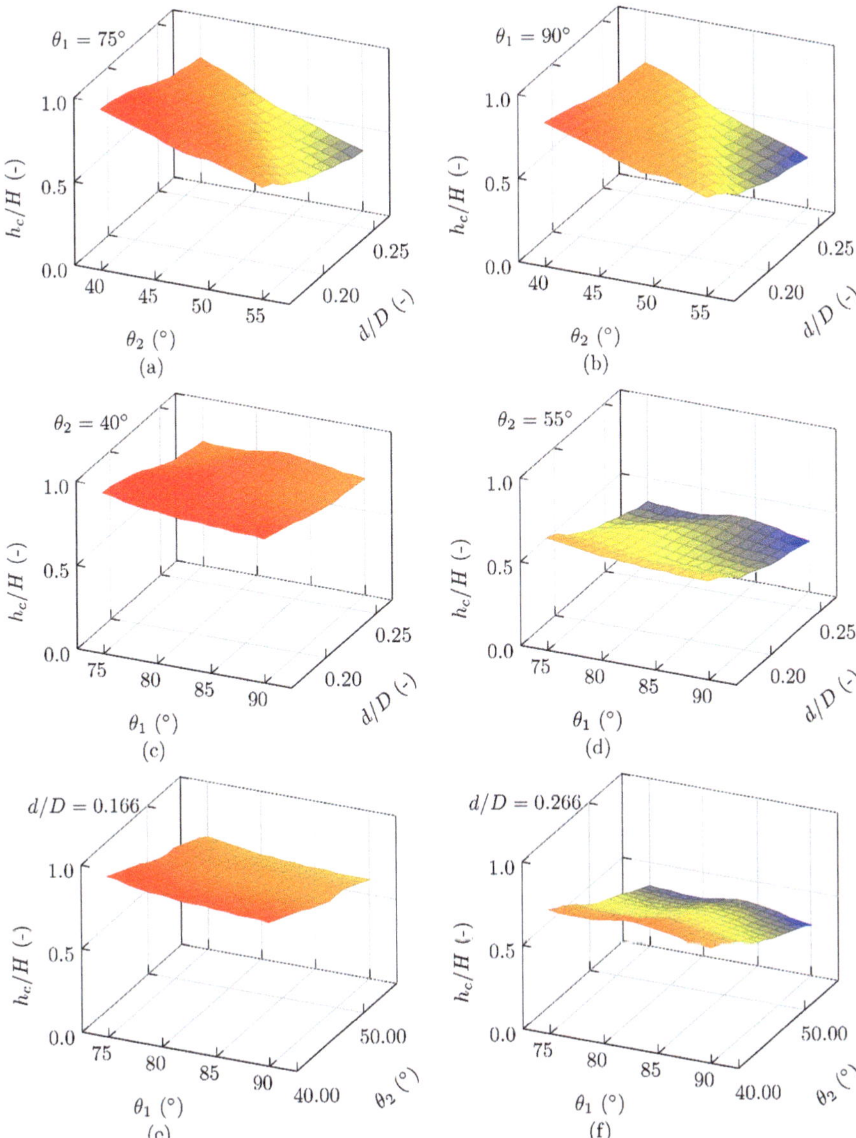

Fig. 13.14 Relation between the ETP height h_c/H and structural parameters [17]: (**a**) $\theta_1 = 75°$, (**b**) $\theta_1 = 90°$, (**c**) $\theta_2 = 40°$, (**d**) $\theta_2 = 55°$, (**e**) $d/D = 0.166$, and (**f**) $d/D = 0.266$

References

1. T.M. Mitchell, *Machine Learning* (McGraw-Hill Education, New York, 1997)
2. Domingos; Pedro., A few useful things to know about machine learning. Commun. ACM **55**(10), 78–87 (2012)
3. S. Bhattacharyya, S. Jha, K. Tharakunnel, J.C. Westland, Data mining for credit card fraud: a comparative study. Decis. Support. Syst. **50**(3), 602–613 (2011)
4. G. Ambrish, B. Ganesh, A. Ganesh, C. Srinivas, K. Mensinkal, Logistic regression technique for prediction of cardiovascular disease. Glob. Trans. Proc. **3**(1), 127–130 (2022)
5. T.M. Jawa, Logistic regression analysis for studying the impact of home quarantine on psychological health during COVID-19 in Saudi Arabia. Alex. Eng. J. **61**(10), 7995–8005 (2022)
6. M. Ihme, W. Tong, A. Ananda, Combustion machine learning: principles , progress and prospects. Prog. Energy Combust. Sci. **2022**(91), 101010 (2021)
7. J. Singh Kushwah, A. Kumar, S. Patel, R. Soni, A. Gawande, S. Gupta, Comparative study of Regressor and classifier with decision tree using modern tools. Mater. Today Proc. **56**, 3571–3576 (2022)
8. L. Breiman, Random forests. Mach. Learn. **45**(1), 5–32 (2001)
9. S.L. Brunton, B.R. Noack, P. Koumoutsakos, Machine learning for fluid mechanics. Annu. Rev. Fluid Mech. **52**, 477–508 (2020)
10. Y. Bai, Z. Jin, Prediction of SARS epidemic by BP neural networks with online prediction strategy. Chaos, Solitons Fractals **26**(2), 559–569 (2005)
11. V. Vapnik, *The Nature of Statistical Learning Theory* (Springer, New York, 1995)
12. N. Cristianini, J. Shawetaylor, *An Introduction to Support Vector Machines* (Cambridge University Press, Cambridge, 2000)
13. Y.-P.P. Chen, E.P. Ivanova, F. Wang, P. Carloni, H.-W. Liu, L.B.T.-C.N.P.I.I. Mander (eds.), *Bioinformatics* (Elsevier, Oxford, 2010), pp. 569–593
14. T. Tada, K. Hitomi, Y. Wu, S.-Y. Kim, H. Yamazaki, K. Ishii, K-mean clustering algorithm for processing signals from compound semiconductor detectors. Nucl. Instrum. Methods Phys. Res. Sect. A Accel. Spectrom. Detect. Assoc. Equip. **659**(1), 242–246 (2011)
15. J.W. Chew, R.A. Cocco, Application of machine learning methods to understand and predict circulating fluidized bed riser flow characteristics. Chem. Eng. Sci. **217**, 115503 (2020)
16. Y. Yan, T. Mattisson, P. Moldenhauer, E.J. Anthony, P.T. Clough, Applying machine learning algorithms in estimating the performance of heterogeneous, multi-component materials as oxygen carriers for chemical-looping processes. Chem. Eng. J. **387**, 124072 (2020)
17. Y.L. Shao, R.K. Agarwal, X.D. Wang, B.S. Jin, Study of flow patterns in a moving bed reactor for chemical looping combustion based on machine learning methods. J. Energy Resour. Technol. ASME **6**, 145 (2023)
18. Y. Shao, R.K. Agarwal, J. Li, X. Wang, B. Jin, Computational fluid dynamics-discrete element model simulation of flow characteristics and solids' residence time distribution in a moving bed air reactor for chemical looping combustion. Ind. Eng. Chem. Res. **59**(40), 18180–18192 (2020)

Chapter 14
Chemical Looping Beyond Combustion

As a rapidly emerging combustion technology for clean and high-efficiency utilization of fossil and renewable fuels, chemical looping combustion (CLC) has received widespread attention, and extensive related research has been conducted. Previous chapters have given a detailed demonstration of process simulation and CFD/CFD-DEM simulation of CLC process in different types of reactor configurations. The chemical looping concept is very attractive due to its inherent ability to achieve separation of products of combustion and other reactions and processes with very low energy penalty. The basic principle behind chemical looping is to split an overall reaction into two sub-reactions occurring in two separate reactors which are integrated using a specific type of carrier which is continuously circulating between the two reactors. When the concept is applied to combustion, it is called the chemical looping combustion, and the circulation material called the oxygen carrier is oxidized by the air in the air reactor also called the oxidizer. In the fuel reactor, the fuels get fully oxidized by obtaining the lattice oxygen from the oxygen carrier producing a gaseous mixture of CO_2 and H_2O.

The initial early interest in chemical looping has been focused on combustion, but recently the concept has been applied to other fuel conversion reactions beyond combustion; these applications are combined together in the literature under the title, "chemical looping beyond combustion (CLBC)." Different from full oxidation of fuels in CLC, the fuels can be partially oxidized via the use of appropriate oxygen carriers or control of carrier residence time and fuel to air ratio. The synthesis gas, namely, CO and H_2 can be obtained after the incomplete combustion. Beyond direct oxidation, CLBC can also be applied to the reforming process in different forms. Steam and CO_2 can also be used to regenerate the reduced oxygen-carrying agent in the redox cycles resulting in the production of H_2 or CO at the same time. Many value-added chemical products such as syngas, hydrogen, alkene, aromatic compounds, etc. can also be created using the CLBC process with high efficiency, high purity, and low energy consumption. Several typical and promising CLBC applications and principles behind them are provided in the following sections.

R. K. Agarwal, Y. Shao, *Modeling and Simulation of Fluidized Bed Reactors for Chemical Looping Combustion*, https://doi.org/10.1007/978-3-031-11335-2_14

14.1 Chemical Looping Air Separation

To realize the simple and efficient oxygen production, chemical looping air separation (CLAS) was proposed by Moghtaderi [1] in 2010. The schematic diagram of this technology is shown in Fig. 14.1. In the reduction reactor, the oxygen carrier releases O_2 in the atmosphere of steam as given in Eq. (14.1), while the reduced oxygen carrier reacts with the oxygen from the air in the oxidization reactor as given in Eq. (14.2). The inert steam can reduce the actual oxygen partial pressure, thereby promoting oxygen release from the oxygen carrier. In the outlet of the reduction reactor, the gas mixture includes the released O_2 and the superheated steam, and the high-concentration O_2 can be obtained via the simple condensation. In the CLAS technology, the carrier gas plays the role of maintaining the low-oxygen partial pressure in the reduction reactor and continuously transporting the generated oxygen out of the reactor. Different types of reactors can be designed for the CLAS system, and two interconnected fluidized bed arrangement is one of them. The reduction reactor and oxidation reactor can be designed as a low-velocity bubbling fluidized bed and a high-velocity circulating fluidized bed riser, respectively. Since the purpose of the CLAS process is to produce oxygen, the capability of oxygen carrier releasing oxygen in reduction reactor is very important. Through thermodynamic analysis, it is found that three types of single metal oxides can be used for the CLAS process which are CuO/Cu_2O, CO_3O_4/CO, and Mn_2O_3/Mn_3O_4:

$$Me_xO_{y-2}(s) \rightarrow Me_xO_y(s) + O_2(g) \tag{14.1}$$

$$Me_xO_{y-2}(s) + O_2(g) \rightarrow Me_xO_y(s) \tag{14.2}$$

Fig. 14.1 Schematic diagram of a CLAS system [2]

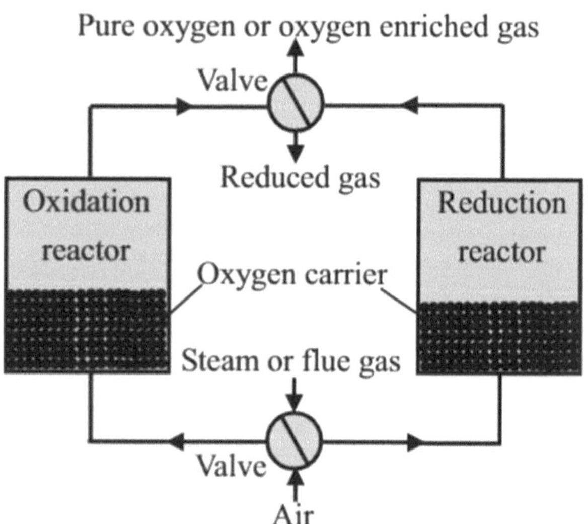

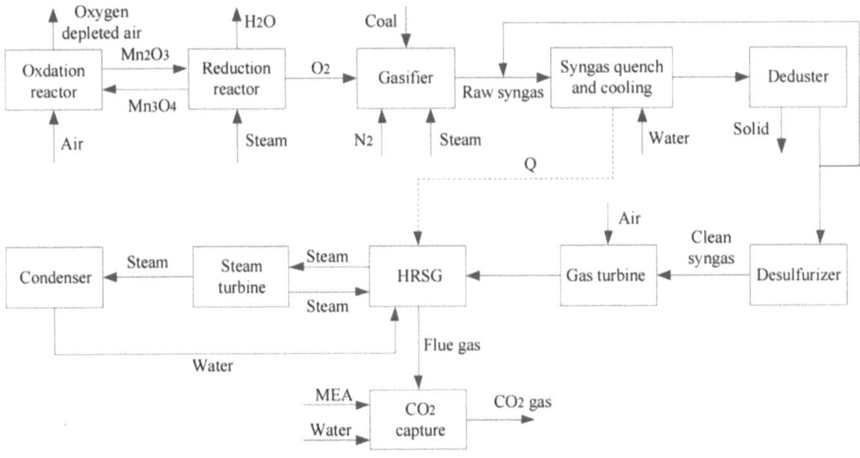

Fig. 14.2 CLAS-IGCC system with post-combustion CO_2 capture [3]

In the CLAS process, when the steam is used as the inert gas, pure oxygen can be obtained, and the gas can be further stored and transported after compression or directly used for the integrated gasification combined cycle. When CO_2 is applied as the inert gas instead, the mixture gas of O_2 and CO_2 at the outlet of reduction reactor can be used for some special combustion conditions such as oxy-fuel combustion. In the perspective of the system efficiency, the CLAS process is very efficient because very little energy is required. From Eq. (14.1) and Eq.(14.2), it can be seen that the net heat released from the CLAS process is zero in theory, which means that the heat released in the oxidation reactor can be sufficient enough to support the endothermic reaction in the reduction reactor. In addition, the heat of the reduced air stream and the superheated steam/high-temperature CO_2 can be used to produce the steam injected to the reduction reactor and preheat the air fed into the oxidation reactor.

Cao et al. [3] combined CLAS technology and integrated gasification combined cycle (IGCC) with CO_2 capture, as shown in Fig. 14.2. In the CLAS subsystem, pure oxygen can be obtained from redox reaction of the oxygen carriers. In the IGCC subsystem, H_2O and oxygen obtained from the CLAS process are injected to the coal gasifier to produce raw syngas rich in H_2 and CO. The product is further cooled, and the sensible heat is recycled. The raw syngas is injected to the deduster to remove the dust and to the desulfurizer to remove H_2S and COS. After that, the clean syngas is fed into the combustion chamber to mix and react with air and do the work in the gas turbine. The extinct heat of the gas turbine flue gas is recovered in the heat recovery steam generator (HRSG) to provide steam for the steam turbine, and the exhausted steam from the steam turbine is finally condensed in the condenser. The CO_2 in flue gas from the HRSG is finally captured using monoethanolamine. In addition, Cao et al. [3] also built a CLAS-IGCC system with pre-combustion CO_2 capture using polyethylene glycol dimethyl ether. The energy losses and exergy destructions as well as the effects of oxygen-to-coal mass ratio, steam-to-coal mass ratio, and temperature in the reduction reactor were investigated in the two systems.

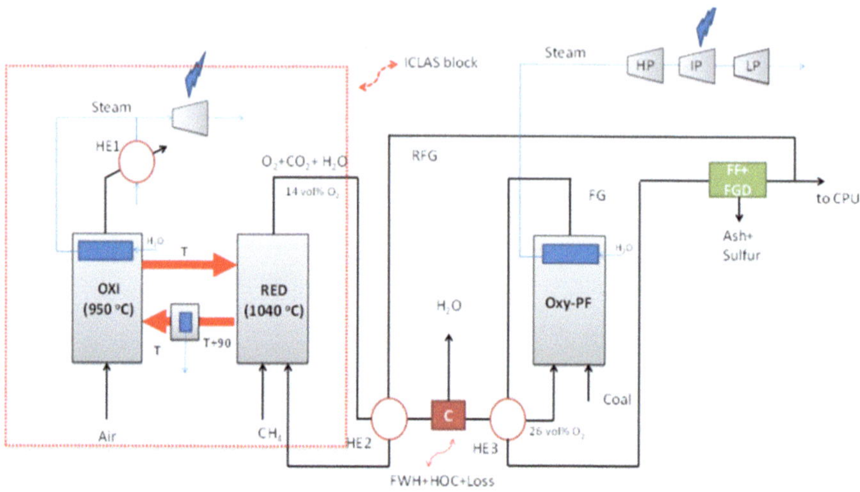

Fig. 14.3 CLAS-FG system with CH_4 firing in the reduction reactor [4]

Zhou et al. [4] combined CLAS process with oxy-fuel combustion. Oxy-fuel combustion has been regarded as one of the most promising low-emission combustion technologies. However, the major bottleneck of commercialization of this technology is the high energy penalties of the air separation process. In the work of Zhou et al. [4] the oxygen required for the oxy-fuel combustion process is produced from the CLAS process as shown in Fig. 14.3. Methane is introduced to the reduction reactor to provide sufficient energy required for the reduction reaction. The oxygen in the reduction reactor outlet is diluted because of the methane firing, and 26 vol % O_2 can be obtained after condensation. The flue gas composed of O_2 and CO_2 is then introduced to the inlet of the oxy-pulverized fuel (PF) firing reactor, while the recycled flue gas (RFG) from the oxy-PF is preheated by the flue gas from the reduction reactor outlet before it enters the reduction reactor. In addition, the CLAS-FG system with indirect methane firing and solar heating with steam as the sweep gas was also investigated. Results confirm the viability of CLAS technology under certain conditions from both technical and economic aspects compared with cryogenic air separation unit which is the only mature technology available for large-scale oxygen production.

14.2   Chemical Looping Reforming

The concept of chemical looping reforming (CLR) was first proposed by Mattisson et al. [5] in 2001. The CLR technology can be categorized into three types which are steam reforming integrated with chemical looping combustion (SR-CLC), autothermal chemical looping reforming (a-CLR), and chemical looping steam methane reforming (CL-SMR).

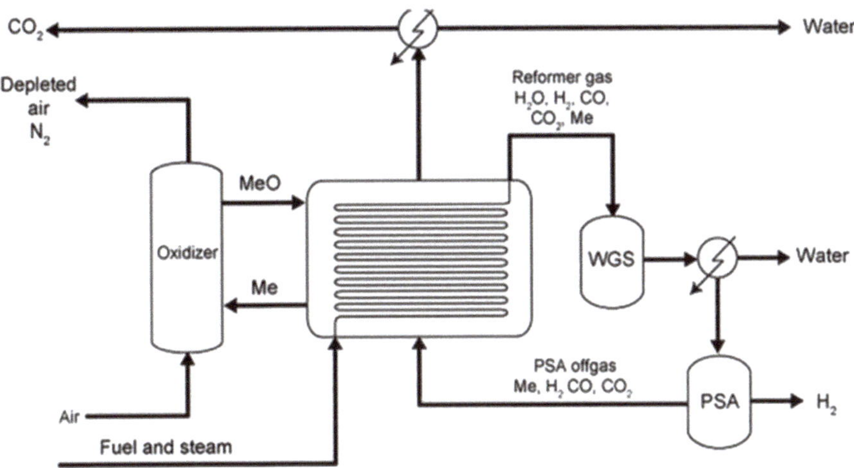

Fig. 14.4   Schematic diagram of SR-CLC [6]

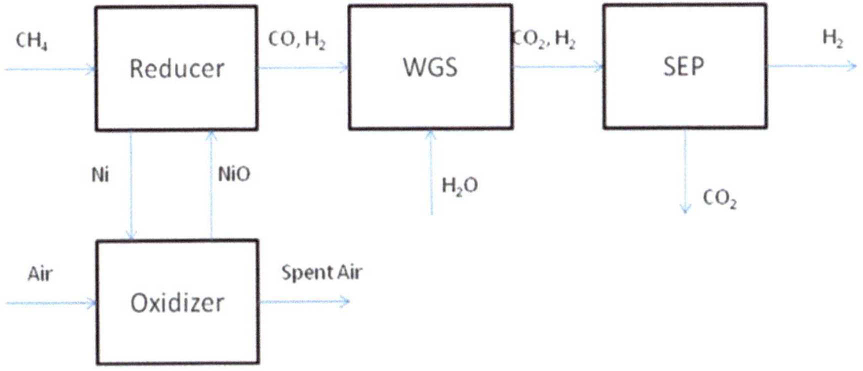

Fig. 14.5   Schematic diagram of a-CLR [6]

The principle of SR-CLC process is illustrated in Fig. 14.4. The steam reforming tubes are placed inside of the reducer. The heat required in the steam reforming process is from the oxygen carrier. This is different from the conventional steam reforming process where the heat is supplied from the fuel combustion outside of the reformer tubes. The reformer gas is transferred to a water-gas shift reactor before it is purified in the pressure swing adsorption (PSA). The pure hydrogen is stored, and the off-gas from the PSA is then fed into the inlet of the reducer. The SR-CLC has the advantages of higher H_2 yield, in situ CO_2 capture, and tail gas utilization. The potential concern of this method is the erosion of the reformer tubes.

The principle of a-CLR is illustrated in Fig. 14.5. Different from the CLC, the mixture gas of H_2 and CO instead of CO_2 and H_2O is produced in the reducer in the a-CLR process. Thus, the air-to-fuel ratio should be kept low to avoid the full oxidization of the fuels. Besides CO and H_2, the flue gas produced from the fuel

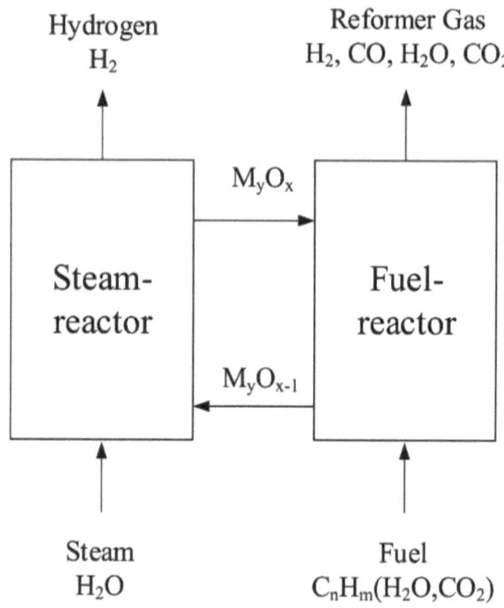

Fig. 14.6 Schematic diagram of CL-SMR [7]

reactor also includes a small amount of CO_2 and H_2O. To enrich the H_2 yield, the mixture gas is further processed in a water-gas-shift (WGS) reactor. The product gas consists of CO_2 and H_2 which is separated via pressure swing adsorption or absorption with suitable amine solvent. For the traditional steam methane reforming process, heat generated from external combustion should be supplied. In the a-CLR process, the circulation particles carry heat from the oxidizer to the reducer to support the gas reforming. Compared with traditional SMR, the H_2 yield can be enhanced in the a-CLR process. However, the a-CLR system does not have the capability of inherent CO_2 separation, and thus the purification of hydrogen is realized via conventional technologies applied to the exit gas.

The CL-SMR is developed to generate pure hydrogen and syngas. As illustrated in Fig. 14.6, in the fuel reactor, the methane gas is injected and partially oxidized by oxygen carriers to produce syngas; in the steam reactor, the reduced oxygen carrier is oxidized by steam to recover lattice oxygen instead of being oxidized by air, and the hydrogen is generated. The reactions in the fuel reactor and steam reactor are given in Eq. (14.3) and Eq. (14.4), respectively. The CL-SMR technology is very promising as it can produce syngas in fuel reactor and H_2 in steam reactor simultaneously without any further energy-consuming gas treatment:

$$Me_yO_x + \delta CH_4 \rightarrow Me_yO_{x-\delta} + \delta(2H_2 + CO) \tag{14.3}$$

$$Me_yO_{x-\delta} + \delta H_2O \rightarrow Me_yO_x + \delta H_2 \tag{14.4}$$

14.3   Chemical Looping Hydrogen Production

In chemical looping hydrogen (CLH) production technology, three reactors are required including an air reactor (AR), a fuel reactor (FR), and a steam reactor (SR), as displayed in Fig. 14.7. When Fe_2O_3 is used as the oxygen carrier circulating between different reactors, the reaction between gaseous fuel CH_4 and oxygen carrier in the fuel reactor is shown in Eq. (14.5). Pure CO_2 can be obtained after condensation in fuel reactor outlet. The reduced oxygen carrier FeO is then transported to the steam reactor where reduced oxygen carriers are partially oxidized to Fe_3O_4 by steam and hydrogen is produced, as given in Eq. (14.6). Further, the partially oxidized oxygen carrier particles are circulated to the air reactor and get oxidized to their original state, as given in Eq. (14.7). Finally, the oxidized oxygen carrier particles move to the fuel reactor to begin another cycle. Besides CH_4, syngas (CO: H_2) can also be used as the reducing agent in the fuel reactor. Overall, the reactions in the steam reactor and air reactor are exothermal, while reactions in the fuel reactor are endothermic, which makes the CLH process a thermally balanced process. Thus, the CLH technology has the advantages of elimination of refining steps producing high-purity H_2 and balancing the whole heat requirements of reactions:

$$Fe_2O_3 + 0.25CH_4 \rightarrow 2FeO + 0.5H_2O + 0.25CO_2 \tag{14.5}$$

$$2FeO + 0.67H_2O \rightarrow 0.67Fe_3O_4 + 0.67H_2 \tag{14.6}$$

$$0.67Fe_3O_4 + 0.17O_2 \rightarrow Fe_2O_3 \tag{14.7}$$

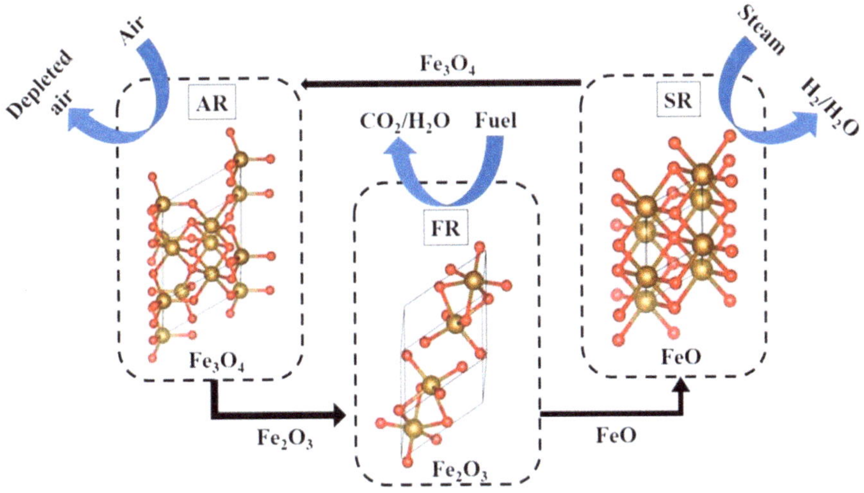

Fig. 14.7 Schematic of chemical looping hydrogen generation process [8]

14.4 Chemical Looping Ammonia Production

The Haber-Bosch (HB) process is widely accepted as one of the most important methods for industrial NH_3 production. It is based on the high-temperature (300–500 °C) and high-pressure (20.26–30.39 MPa) reaction between N_2 and H_2 in the presence of a suitable catalyst, as shown in Eq. (14.8). In this process, a larger amount of energy is required, and the conversion rate is only about 10–15%. In addition, 1–2.5% of annual CO_2 emissions are from the use of natural gas as a hydrogen source. Chemical looping ammonia production (CLAP) is an innovative and environmentally friendly low-pressure ammonia synthesis technology. It separates the HB process into two sub-reactions, namely, nitridation step and ammonia synthesis step, which are connected by nitrogen carriers. Nitrogen carriers can be regarded as a nitrogen storage medium for activated nitrogen used for charging and discharging nitride lattices, which is similar as the oxygen carrier used for chemical looping combustion process. Figure 14.8 illustrates schematic diagram of CLAP process. In the ammonia production reactor, nitrogen carrier reacts with H_2 or H_2O to form NH_3, while in the nitridation reactor, the N-depleted nitrogen carrier is regenerated by N_2 or N_2 with reducing agent. The regenerated nitrogen carriers are then sent back to the ammonia production reactor to start a new reaction cycle. Compared with conventional HB process, CLAP can prevent the competitive adsorption of N_2 and hydrogen sources (H_2 or H_2O) on the surface of catalyst. In the HB process, the thermodynamic and kinetic properties are contradictory in certain conditions, as the reaction rate is decreased in low temperature, but high temperature is not beneficial for the exothermic reaction. CLAP process can optimize the thermodynamic and kinetic properties at the same time by splitting the overall reaction into two sub-processes occurring in two separate reactors. Relatively high temperature in nitridation reactor can break the $N{\equiv}N$ triple bond with high thermodynamic strength, ensuring high nitridation efficiency. The nitrogen release process has low-temperature requirements, and relatively low temperature in ammonia production reactor can ease the decomposition of NH_3, which is conducive to the collection of NH_3:

$$N_2 + H_2 \rightarrow NH_3 \quad \Delta H = -92.4 \ \text{kJ/mol} \tag{14.8}$$

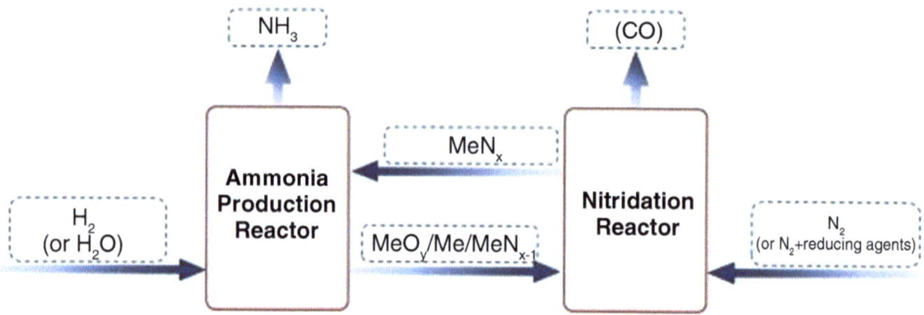

Fig. 14.8 Schematic diagram of chemical looping ammonia production process [9]

14.5   Chemical Looping Oxidative Coupling of Methane

The oxidative coupling of methane (OCM) is a method to realize the direct conversion of CH_4 to ethane and ethylene at temperature of about 700–900 °C, which is expressed by the reaction in Eq. (14.9) and Eq. (14.10). The OCM reaction was proposed by Keller and Bhasin in 1982 based on the oxygen and methane co-feed route [10]. Various promising catalysts have been shown to be useful for OCM such as transition metal oxides, composite oxides, alkali and alkaline earth-promoted metal oxides, etc. However, existing studies point out that it is rather difficult to achieve the C_2 yields above 25–30%, which is the main reason why OCM has not come into industrial application. Much research points out that a heterogeneous-homogeneous reaction step is included in OCM. In the surface of the catalysts, the CH_4 molecule is activated into CH_3 radicals on the electrophilic oxygen site, and then a combination of two CH_3 radicals forms C_2H_6. Furthermore, two hydrogen ions are dehydrogenated from C_2H_6 to generate C_2H_4 molecule. The explanation of active site for CH_4 activation and C_2 hydrocarbon formation is still controversial, but it is widely accepted that the surface active oxygen species of catalysts are of great importance for OCM reactions. In the traditional OCM method, the existence of gaseous oxygen can promote the conversion of total/part of hydrocarbon to CO or CO_2, which is not expected as the main goal of OCM is to achieve high C_2 selectivity. In addition, pure oxygen is needed for OCM, which requires addition of uneconomic air separation process. Thus, it is of great importance to have an alternative treatment option which can realize the conversion of CH_4 to C_2 hydrocarbons without the demand of pure oxygen:

$$4CH_4 + O_2 \rightarrow 2C_2H_6 + 2H_2O \ \Delta H = -176.6 \ kJ/mol \tag{14.9}$$

$$2CH_4 + O_2 \rightarrow C_2H_4 + 2H_2O \ \Delta H = -105.5 \ kJ/mol \tag{14.10}$$

Chemical looping oxidative coupling of methane (CL-OCM) is an alternative technology to conventional OCM. Figure 14.9 displays the principle of CL-OCM. In this method, the methane is oxidized in the reduction reactor by the metal oxides to form coupling products. Similar to the CLC or CLOU process, the reduced oxygen carrier then flows to the oxidation reactor where particles get contact and react with the air to get re-oxidized and release heat. The CL-OCM technology has some advantages over the conventional OCM treatment method. C_2 selectivity in CL-OCM can be improved due to the fact that methane is not exposed to the molecular oxygen and the lattice oxygen from the metal oxides is used instead. The upper yield limit for C_2 products can be improved by about 60% at elevated pressure in the absence of gas phase reaction via the CL-OCM process.

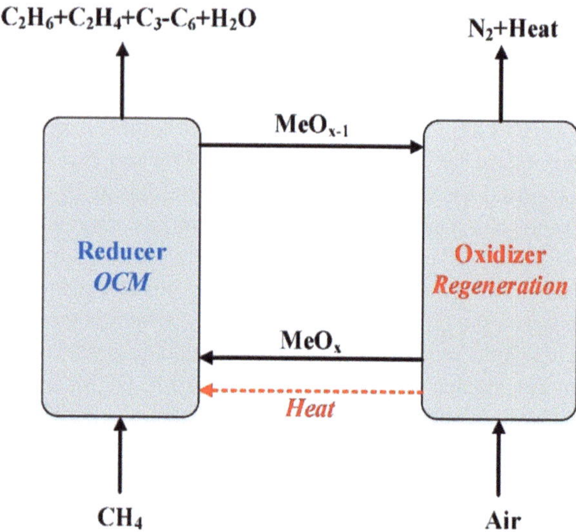

Fig. 14.9 Schematic diagram of chemical looping oxidative coupling of methane [11]

14.6 Chemical Looping Dehydroaromatization

The methane dehydroaromatization (DHA) reaction is a desirable process to directly convert methane to liquid aromatics in only one step. It was first reported in 1993 by passing methane over a ZSM-5 zeolite modified with Mo or Zn. In the methane DHA process, no other reactant besides methane is needed, and it is usually conducted at 1 atm. Without requiring high operating pressures. In spite of this, the methane DHA has not yet been commercialized, and the following two main obstacles should be overcome. First, limited by the thermodynamic equilibrium the yield of benzene is still low, which is only about 12–13% at 700 °C. Secondly, the decomposition of CH_4 inevitably happens under specified conditions, and the coke is accumulated on the surface of the catalysts which account for 10–20% of methane conversion. This further causes the deactivation of the catalysts after a few hours.

The chemical looping dehydroaromatization (CL-DHA) process proposed by Brady et al. [12] includes four steps. In the first step, the DHA reaction happens at the 700 °C to produce a mixture of methane, aromatics, and H_2. In step 2, H_2 from the DHA effluent is oxidized by the metal oxide such as Fe_3O_4 to form H_2O. In step 3, the reduced oxide from step 2 reacts with steam to realize oxidization and produce H_2. In step 4, the high-temperature steam in step 2 is removed from the product gas using the water sorbent which can be regenerated via a temperature swing. The above steps realize the separation of H_2 from the DHA gas stream, and the final gases only include CH_4 and C_6H_6. The mixture gas is further fed into the DHA unit to act as the thermodynamic driving force for DHA reaction. In theory, the CL-DHA process is able to achieve 100% atom efficiency to create valuable products including aromatics and H_2.

14.7   Chemical Looping Oxidative Dehydrogenation

Ethane oxidative dehydrogenation (ODH) is an alternative to steam cracking for ethylene production. In this approach, ethane and gaseous oxygen are directly converted into ethylene and water in the presence of a heterogeneous catalyst, as given in Eq. (14.11). The hydrogen is oxidized by oxygen, and thus the ethylene conversion is not subject to the equilibrium limitations. The overall reaction is exothermic with $\Delta H = -105$ kJ/mol at 850 °C. However, there are some disadvantages in the ODH process. The use of oxygen as the reactant requires an air separation unit. To ensure safety, the ethylene partial pressure in the mixture gas should be kept low which is usually smaller than 0.1 bars. In addition, high per pass ethylene yield need to be obtained in view of the high complexity and cost of oxygenated byproducts removal, but catalysts with high activity and selectivity are still currently lacking:

$$2C_2H_6 + O_2 \rightarrow 2C_2H_4 + 2H_2O \tag{14.11}$$

To address the above challenges, the chemical looping oxidative dehydrogenation (CL-ODH) has been proposed. As illustrated in Fig. 14.10, the ODH process occurs in the reducer where the ethane reacts with the lattice oxygen provided from a redox catalyst (i.e., oxygen carrier); the reduced catalysts are then returned to the oxidizer to react with air. Compared to ODH, the CL-ODH process avoids direct contact of ethane and gaseous oxygen, which removes the safety concern and does not require a large fraction of inert dilution gas to maintain the low ethane partial pressure. The oxygen is transferred from air by redox catalysts, and air separation unit is not needed in the two-step process. Superior olefin yields higher than 70% have been realized at the ethane partial pressure over 0.8 bars in CL-ODH. High ethylene yield and the integrated oxygen separation reduce the load for the

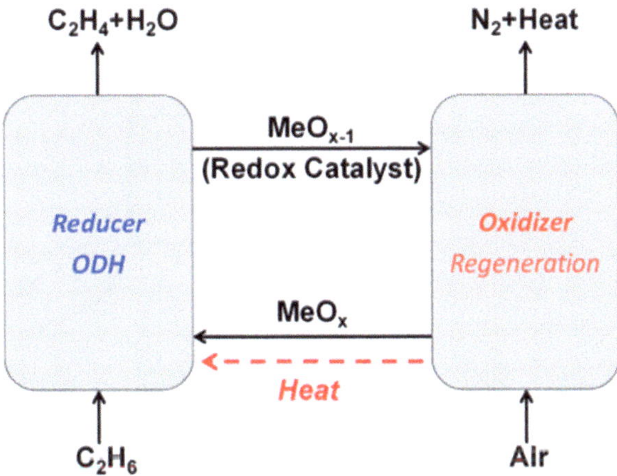

Fig. 14.10 Schematic diagram of CL-ODH [13]

downstream purification steps. In theory, higher ethylene selectively is promising to be obtained when the redox catalyst used in the two-step cycles is properly designed. Besides ethane, CL-ODH strategy can also be applied to the conversion of other light alkanes.

14.8   Chemical Looping Reverse Water-Gas Shift

Reverse water-gas shift (RWGS) is the reaction transforming the CO_2 and H_2 to CO and H_2O under relatively mild conditions as given in Eq. (14.12). Following the reduction of CO_2, the generated CO can be further used as the feedstock for the production of some energy-dense liquid hydrocarbons such as methanol by Fischer-Tropsch synthesis (FTS). Nevertheless, the reaction rate is inhibited by the water produced in RWGS, and the catalysts could be deactivated during methanol generation via FTS. Thus, condensation of the mixture gas is required, but this further decreases the utilization efficiency of the heat energy at high temperature:

$$CO_2 + H_2 \rightarrow CO + H_2O \quad \Delta H^0 = 41.33 \text{ kJ/mol} \quad (14.12)$$

The reverse water-gas shift chemical looping (RWGS-CL) reaction is a two-step process that realizes the conversion of CO_2 to CO. As illustrated in Fig. 14.11, in the reduction step, H_2 reacts with the lattice oxygen from the oxygen carrier to form H_2O. In the oxidation step, the reduced oxygen carriers get contact with CO_2. The oxygen vacancies on the particle surface act as the active sites which reduce the injected CO_2 to CO and obtain oxygen to realize regeneration at the same time. In the RWGS-CL reaction scheme, the reactions of CO_2 and H_2 occur in separate reactors. The avoidance of direct contact prevents the formation of undesirable byproducts. The generated CO is not mixed with steam, so the high-purity gas stream does not

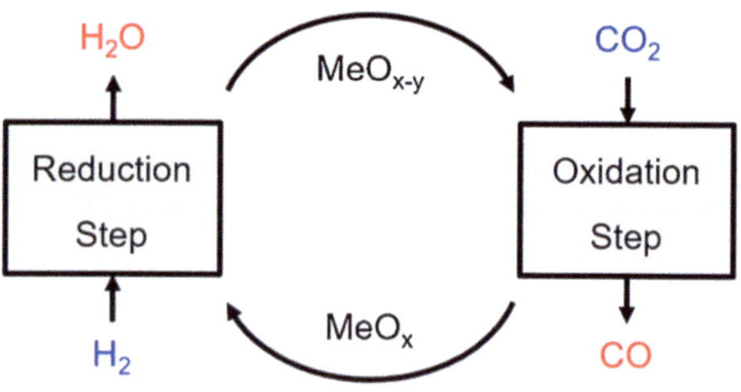

Fig. 14.11 Schematic diagram of RWGS-CL [14]

Reduction step : $MeO_x + yH_2 \rightarrow MeO_{x-y} + yH_2O$

Oxidation step : $MeO_{x-y} + yCO_2 \rightarrow MeO_x + yCO$

require the water removal process anymore. In addition, the excess amount of H_2 is needed for kinetic and thermodynamic reasons in traditional RWGS reaction, while H_2 use is minimized in RWGS-CL process, which is more cost-saving and energy-efficient.

This chapter has provided several key examples of application of chemical looping concept beyond combustion. One can find many more examples in the literature.

References

1. B. Moghtaderi, Application of chemical looping concept for air separation at high temperatures. Energy Fuel **24**(1), 190–198 (2010)
2. K. Wang, Q. Yu, Q. Qin, W. Duan, Feasibility of a co oxygen carrier for chemical looping air separation: Thermodynamics and kinetics. Chem. Eng. Technol. **37**(9), 1500–1506 (2014)
3. Y. Cao, B. He, G. Ding, L. Su, Z. Duan, Energy and exergy investigation on two improved IGCC power plants with different CO_2 capture schemes. Energy **140**, 47–57 (2017)
4. C. Zhou, K. Shah, H. Song, J. Zanganeh, E. Doroodchi, B. Moghtaderi, Integration options and economic analysis of an integrated chemical looping air separation process for oxy-fuel combustion. Energy Fuels **30**(3), 1741–1755 (2016)
5. T. Mattisson, A. Lyngfelt, Applications of chemical-looping combustion with capture of CO_2. *Second Nord. Minisymp. CO_2 Capture Storage, Göteborg* (Sweden, 2001)
6. L.S. Fan, L. Zeng, W. Wang, S. Luo, Chemical looping processes for CO_2 capture and carbonaceous fuel conversion – prospect and opportunity. Energy Environ. Sci. **5**(6), 7254–7280 (2012)
7. M. Luo, Y. Yi, S. Wang, Z. Wang, M. Du, J. Pan, Q. Wang, Review of hydrogen production using chemical-looping technology. Renew. Sust. Energ. Rev. **81**, 3186–3214 (2018)
8. S. Das, A. Biswas, C.S. Tiwary, M. Paliwal, Hydrogen production using chemical looping technology: A review with emphasis on H2 yield of various oxygen carriers. Int. J. Hydrog. Energy **47**(66), 28322–28352 (2022)
9. S. Yang, T. Zhang, Y. Yang, B. Wang, J. Li, Z. Gong, Z. Yao, W. Du, S. Liu, Z. Yu, Molybdenum-based nitrogen carrier for ammonia production via a chemical looping route. Appl. Catal. B Environ. **312**, 121404 (2022)
10. G.E. Keller, M.M. Bhasin, Synthesis of ethylene via oxidative coupling of methane: I. Determination of active catalysts. J. Catal. **73**(1), 9–19 (1982)
11. S. Jiang, W. Ding, K. Zhao, Z. Huang, G. Wei, Y. Feng, Y. Lv, F. He, Enhanced chemical looping oxidative coupling of methane by Na-doped LaMnO3 redox catalysts. Fuel **299**, 120932 (2021)
12. C. Brady, B. Murphy, B. Xu, Enhanced methane dehydroaromatization via coupling with chemical looping. ACS Catal. **7**(6), 3924–3928 (2017)
13. X. Zhu, Q. Imtiaz, F. Donat, C.R. Müller, F. Li, Chemical looping beyond combustion-a perspective. Energy Environ. Sci. **13**(3), 772–804 (2020)
14. A. Jo, Y. Kim, H.S. Lim, M. Lee, D. Kang, J.W. Lee, Controlled template removal from Nanocast La0.8Sr0.2FeO3 for enhanced CO_2 conversion by reverse water gas shift chemical looping. J. CO2 Util. **56**, 101845 (2022)

Chapter 15
Chemical Looping Combined with Carbon Capture and Sequestration

Geological carbon sequestration (GCS) can provide a quick, efficient, and economical solution to excessive anthropogenic carbon emission without drastic change in energy-generating sources and technologies. Various geological structures have been identified for possible deployment of GCS—deep saline aquifers, depleted oil/gas reservoirs, un-mineable coal seams, etc. According to the estimates by the US Energy Information Administration, deep saline aquifers appear to be the most viable candidates since their storage potential is sufficiently large to achieve the required carbon emission reduction target. Geological surveys and pilot studies of saline aquifer geological carbon sequestration (SAGCS) can be dated back to 1990s. Although some promising results have been obtained, this technology is still not considered mature for large-scale industrial deployment since many uncertainties about sequestration efficiency and safety still exist.

The large spatial extent of the order of kilometers and time duration of the order of centuries for CO_2 plume migration after injection makes the study of SAGCS very difficult at these large spatial and temporal macroscales by using laboratory scale experiments, which can be conducted only on relatively small spatial and temporal scales varying from nanometers to a few meters and from nanoseconds to a few days/ months. Conducting field tests in large-scale formations before the actual deployment takes place can be very expensive. However, numerical simulations using computational fluid dynamics (CFD) technology can be employed at industrial scale to determine the fate of injected CO_2 in a reservoir. With the proper modeling of the storage formation and groundwater transportation, CFD can provide accurate enough analysis for quick estimation of reservoir performance at a considerably lower cost. The governing equations of mass/energy transport and numerical representations of the formation properties have been well explained in the TOUGH2 User's Guide. Another important benefit of numerical simulations is that one can investigate the effect of various injection parameters such as injection rate, injection duration, and injection well orientation and displacement on CO_2 storage efficiency and plume migration in a given reservoir.

R. K. Agarwal, Y. Shao, *Modeling and Simulation of Fluidized Bed Reactors for Chemical Looping Combustion*, https://doi.org/10.1007/978-3-031-11335-2_15

The accurate large-scale simulations of existing (completed/continuing) SAGCS projects for identified known aquifers are crucial for creating confidence in the future deployment of SAGCS projects. Although detailed history-matching simulations of existing SAGCS projects are challenging due to various uncertainties, e.g., in the reservoir topography and hydrogeology, the simulations can still provide informative insights into several aspects of SAGCS, such as the variance in multiphase flow properties, integrity of the geological seals, and the mechanism of CO_2 trapping. Such insights are essential for better understanding of the nature of SAGCS and its best practices for deployment. Detailed history-matching simulations have always been an important part in the SAGCS research activity.

15.1 SAGCS Simulation for Mt. Simon Formation

Located in the Illinois basin, Mt. Simon sandstone formation is a huge saline aquifer that covers most of Illinois, southwestern Indiana, southern Ohio, and western Kentucky. The estimated storage capacity of Mt. Simon formation ranges from 27,000 to 109,000 × 10^6 tons of CO_2 [1]. With the consideration of data availability, a candidate site for future sequestration project, the Weaber-Horn #1 well, has been chosen for the simulation study.

A cylindrical model of Mt. Simon formation is constructed. For thermal condition, the model uses calculated values with a thermal gradient of 9.2 °C/km. The reservoir pressure is assumed to be hydrostatic pressure with a gradient of about 10.8 MPa/km from the ground surface. Salinity is assumed to increase with the depth, starting from 235 mg/L at 450 m below ground surface with a gradient of 12.8 mg/L per meter in depth. A north-south geological slope of 8 m/km is also considered in the modeling. "No-flux" boundary condition is applied at top and bottom of the model, representing the impermeable upper and lower bounding formations. "Fixed-state" boundary condition is imposed at the lateral boundary to represent an essentially "open" reservoir. The permeability and porosity of the 24 sublayers can be seen in Fig. 15.1.

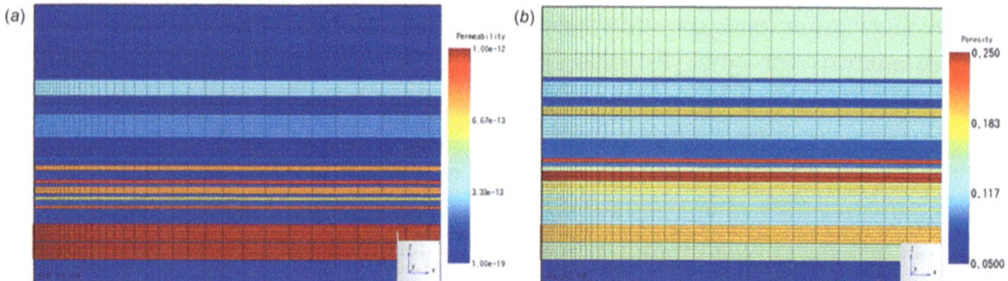

Fig. 15.1 (**a**) Permeability, (**b**) porosity, and computational mesh of the 24 sublayers of the Mt. Simon formation model at WH #1 well [2]

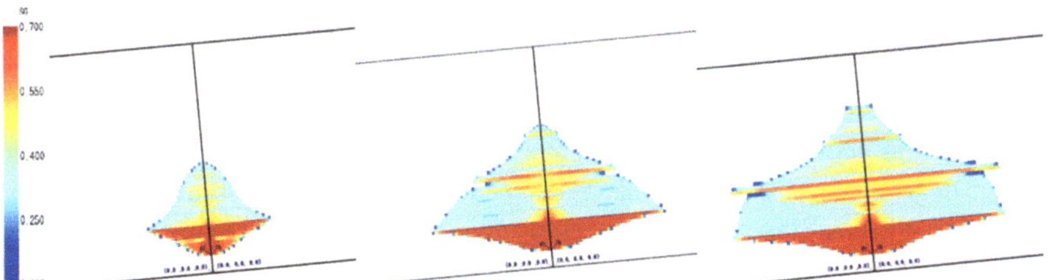

Fig. 15.2 Saturation of gaseous CO$_2$ at (**a**) 5th, (**b**) 25th, and (**c**) 50th year of injection [3]

Due to the relatively high porosity and permeability, CO$_2$ injection is assigned at the bottom of Arkosic unit (bottom three sublayers). The injection rate is assigned to be 5×10^6 tons per year and injection lasts for 50 years. CO$_2$ footprint at 5th year, 25th year, and 50th year since the beginning of injection is examined. Figure 15.2 shows the saturation of gaseous CO$_2$ at (a) 5th, (b) 25th, and (c) 50th years of injection. As seen in Fig. 15.2, CO$_2$ plume evolves with a complex spatial pattern during the 50 years of injection. Within the Arkosic unit where the injector is located, extensive lateral migration with relatively higher concentration of gaseous CO$_2$ is observed. In the overlying sublayers, however, strong secondary sealing effect that retards the vertical migration of gaseous CO$_2$ is observed as the pyramid-shaped subplume.

15.2   SAGCS Simulation for Utsira Formation

The Sleipner project near the Norwegian coast at North Sea is probably the most well-known, important, and successful SAGCS demonstration project so far. It has the most complete topographic description, industrial-scale injection amount, and long-term monitoring data. Nevertheless, great uncertainties still exist for accurate reservoir-scale simulation of the Sleipner SAGCS project. Simulation studies of this project can provide helpful insights in understanding the transport behavior of in situ CO$_2$ and the reservoir performance.

Two numerical models have been constructed for the study of Sleipner SAGCS project. The first model is a generalized axisymmetric layered model for estimating the ballpark migration of in situ CO$_2$. The purpose of this simulation is to determine the secondary sealing effect and gain an overview of the plume migration within Utsira formation. The second model describes a total of 48 km^2 area of detailed topmost sandstone layer (marked as layer #9 in 10). Layer #9 is of particular interest regarding the safety of the sequestration project, as it is the layer within which highest concentration of gaseous CO$_2$ exists and most significant plume migration occurs. Detailed topography of layer #9 is shown in model #2, making it a complicated three dimensional (3D) model. The 3D layer #9 model is introduced to investigate the effect of actual topography on in situ CO$_2$ migration, while avoiding intensive computational effort associated with full 3D modeling and simulation of the entire Utsira formation.

15.2.1 Model #1: Generalized Stratified Model of Utsira Formation

Pre-injection geological survey has unveiled the layered structure of Utsira formation. The majority of Utsira formation can be identified as an eight-layered structure; however, one extra layer needs to be added to the structure near the injection site due to the existence of a sand wedge. Therefore, a cylindrical domain with nine alternating sandstone and shale layers is constructed as shown in Fig. 15.3. According to the seismic survey, it is assumed that all four shale layers have identical thickness of 5 m, four shallower sandstone layers have identical thickness of 25 m, and the bottom sandstone layer has a thickness of 60 m. It adds up to a total 180 m thickness for the modeled Utsira formation. Lateral radius of the generalized cylindrical model reaches 100 km, which is about the same as the actual extent of the southern part of Utsira formation. According to Audigane et al. [4], all sandstone layers have identical and isotropic hydrogeological properties, and so do the shale layers. Figure 15.3 shows the layered structure and computational mesh of the modeled Utsira formation as well as the location of CO_2 injection. The permeability and porosity are 3 Darcy and 0.42 for sand layer, and 10 mDarcy and 0.1025 for shale layer. Temperature of 37 °C and pressure of 11 MPa is applied as the pre-injection conditions. Van Genuchten-Mualem functions are used to describe both relative permeability and capillary pressure. CO_2 injection of 30 kg/s is assigned as a point source at the middle of the bottom-most sand layer.

The simulation time is set at 15 years, and CO_2 plume profile is examined for each year. Figure 15.4 shows the cross-sectional view of gaseous CO_2 in the reservoir for 10 consecutive years since the inception of the injection. Results shown in Fig. 15.4 provide evidence of strong secondary sealing effect for migration of in situ CO_2. Similar to the case of Mt. Simon SAGCS, the injected CO_2 first migrates upward driven by buoyancy until it gets in contact with the first shale layer. Due to the low permeability and high capillary entry pressure, CO_2 is confined by this shale layer and is forced to migrate radially. Simultaneously, CO_2 concentration builds up beneath the shale layer and finally breaks through the capillary barrier upon sustaining sufficient CO_2 column height. The accumulation-penetration breakthrough takes place each time the CO_2 plume encounters a new shale layer and forms an upside-down pyramid-shaped subplume as documented clearly by the first- and second-year

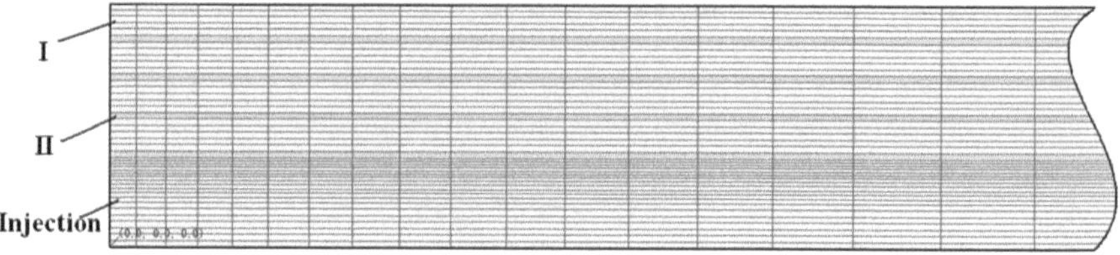

Fig. 15.3 Computational mesh and layered structure of the generalized nine-layered model of Utsira formation [2]

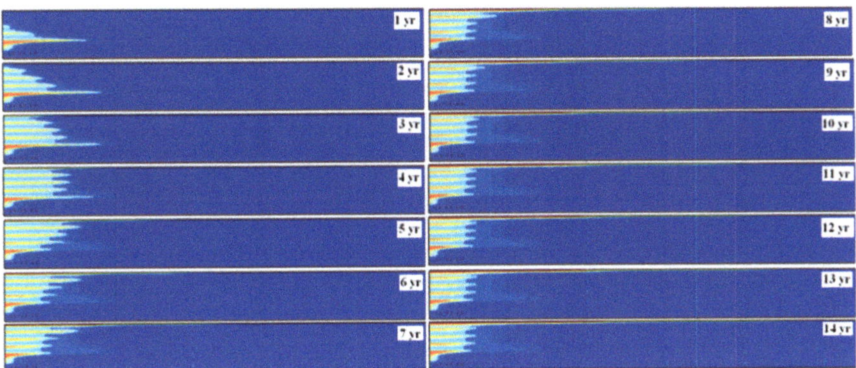

Fig. 15.4 In situ CO_2 distribution for 15 years of injection in Utsira formation model [2]

plume shapes in Fig. 15.4. Due to the secondary sealing effect, in situ CO_2 has very limited contact with the caprock of Utsira formation by the third year of injection.

15.2.2 Model #2: Detailed Three-Dimensional Model of Utsira Layer #9 Formation

In order to examine the plume evolution within the topmost layer more closely, a 3D model of Utsira layer #9 is created with detailed topography. It should be noted that only layer #9, not the entire depth, is modeled because of the following considerations. To ensure the accurate capture of topographic effect on plume shaping, computational domain with considerable fine mesh resolution has to be modeled based on geological survey data. The computational effort and thus the feasibility of highly detailed model of the entire Utsira formation are very intensive and time consuming. Second, CO_2 has to breakthrough several layers of relatively low permeability shale prior to reaching the topmost layer. While it is difficult to quantify the breakthrough of gaseous CO_2, the quantification of CO_2 feeding into the topmost layer (layer #9) is rather reliable. Therefore, a model of only the topmost layer (layer #9) could provide an ideal platform to investigate the effect of various parameters such as topography on the shaping of CO_2 plume, as well as it could be used for optimization purpose to achieve high-efficiency sequestration while maintaining an affordable computational effort and cost.

A reservoir model with dimension of 1600 m × 4900 m with varying thickness is constructed. The topography of this portion of Utsira formation is accurately modeled based on seismic geological survey data (provided by Zhu and Lu of the University of Indiana University [5, 6]) with 50 m × 50 m mesh resolution. Because only layer #9 is modeled, the thickness of the computational domain varies from 3.5 m to 26.3 m with an average thickness of 11.3 m. However, to accurately capture the accumulation and upward and lateral movement of CO_2, 37 layers are used along

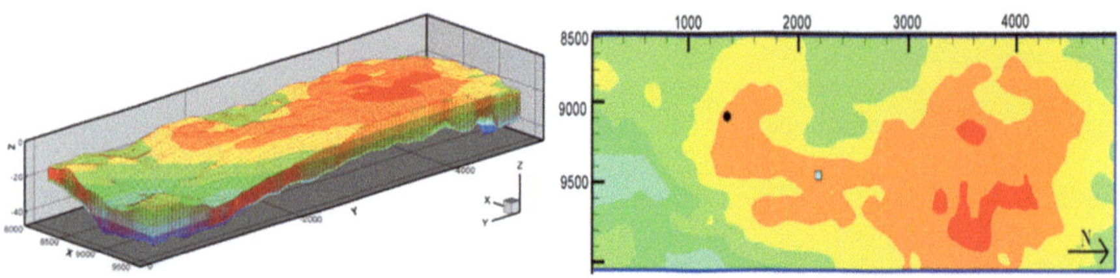

Fig. 15.5 Three-dimensional overview and plan-view of the 3D layer #9 model of Utsira indicating feeder locations (black dot: main feeder; cyan square: secondary feeder) [5]

Table 15.1 Hydrogeological properties of the Utsira layer #9 model [5]

Temperature	33 °C
Pressure	8.6 MPa
Total Utsira formation area	26,100 km^2
Total Utsira formation thickness	50–300 m
Layer #9 area	1600 m × 4900 m
Layer #9 thickness	3.5–26.3 m
Shale permeability	W-E: 0.001 mDarcy, N-S: 0.001 mDarcy, vertical: 0.0001 mDarcy
Mudstone permeability	W-E: 2 Darcy, N-S: 10 Darcy, vertical: 200 mDarcy
Utsira porosity (shale/ mudstone)	35.7%
Residual CO_2 saturation	0.02
Residual brine saturation	0.11
Relative permeability type	Corey/van Genuchten–Muller
Capillary pressure	None
Porewater salinity	3.3%
North-south geological slope	8.2 m/km, 5.8 m/km
CO_2 feeder location	Main feeder: W-E: 516 m, N-S: 1210 m, bottom mudstone Secondary feeder: W-E: 925 m, N-S: 2250 m, bottom mudstone

the thickness. The topmost layer and the bottom two layers are designated to represent the low permeability shale, while the 34 layers in the middle are assigned the properties of mudstone. In the 3D layer #nine model, permeability anisotropy is considered with west-east permeability of 2 Darcy, north-south permeability of 10 Darcy, and vertical permeability of 200 mDarcy. 3D overview of the layer #9 model is shown in Fig. 15.5. Table 15.1 summarizes the hydrogeological properties of the layer #9 model.

The simulation time is set at 9 years, which corresponds to the injection period of 1999–2008. CO_2 plume migration at the topmost layer is examined for each year. Considering all the uncertainties mentioned above, a total of nine simulations are performed until a good history matching is obtained, as shown in Fig. 15.6.

Both the two-dimensional generalized Utsira formation model and the three-dimensional detailed Utsira layer #9 model have generated satisfactory simulation

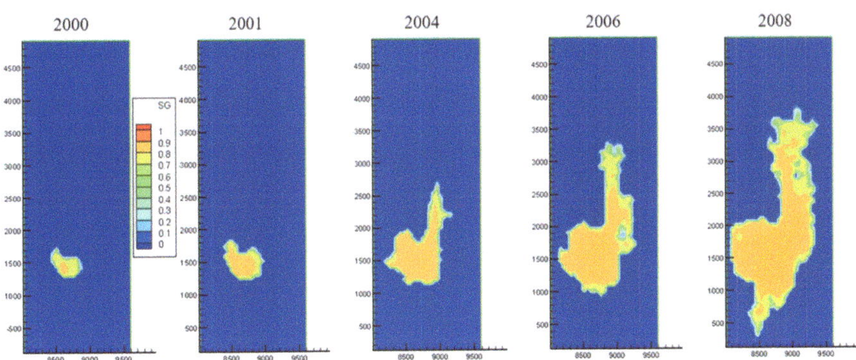

Fig. 15.6 Simulated CO_2 migration in layer #9 of Utsira formation, 2000–2008 [2]

results as seen by history matching. In summary, five major conclusions can be made as follows: first, the simulations show that the permeability anisotropy should be accurately modeled. Vertical-to-horizontal anisotropy close to 10:1 is needed to accurately capture the upward migration of CO_2. Horizontal anisotropy of 2:10 is needed to capture the northern spill of CO_2 into the north-tending ridge. Second, a secondary feeder is likely to exist directly under the north-tending ridge to generate sufficient plume migration along the ridge. It suggests multiple pathways for CO_2 breakthrough from the lower aquifer structure. Third, the fact that injection gas being CO_2-methane mixture is very important in modeling since the presence of 2% methane enhances the buoyancy. Fourth, it is critical that time-dependent feeding of CO_2 is modeled. This is consistent with the behavior of CO_2 path flow breaking the capillary pressure barrier, as noted before for the secondary sealing effect in the case of Mt. Simon formation. Finally, simulation results suggest strong mobility of gaseous CO_2 under the caprock (shale) without major leakage, implying that the caprock serves quite well as the nonpermeable CO_2 barrier while exerting little resistance for the lateral flow of CO_2 underneath.

References

1. D.A. Barnes, D.H. Bacon, S.R. Kelley, Geological sequestration of carbon dioxide in the Cambrian mount Simon sandstone: Regional storage capacity, site characterization, and large-scale injection feasibility, Michigan Basin. Environ. Geosci. **16**(3), 163–183 (2009)
2. Z. Zhang, *Numerical Simulation and Optimization of Carbon Dioxide Sequestration in Saline Aquifers* (Washington University, St Louis, 2013)
3. Q. Zhou, J.T. Birkholzer, E. Mehnert, Y. Lin, K. Zhang, Modeling Basin-and plume-scale processes of CO_2 storage for full-scale deployment. Groundwater **48**(4), 494–514 (2010)

4. P. Audigane, I. Gaus, I. Czernichowski-Lauriol, K. Pruess, T. Xu, Two-dimensional reactive transport modeling of CO_2 injection in a saline aquifer at the Sleipner site, North Sea. Am. J. Sci. **307**(7), 974–1008 (2007)
5. C. Zhu, P. Lu, *Personal Communication; Department of Geological Sciences* (University of Indiana, Bloomington, 2012)
6. Orr, L. Carbon capture and sequestration: Where do we stand? in *Presentation at NAE/AAES Convocation* (Washington, 2010)

Index